Structure and Dynamics of the Earth's Interior 1

SCIENCES

Geoscience, Field Director – Yves Lagabrielle

Earth's Interior, Subject Head – Nathalie Bolfan-Casanova

Structure and Dynamics of the Earth's Interior 1

Dynamics of the Earth's Mantle

Coordinated by
Julien Monteux

iSTE

WILEY

First published 2025 in Great Britain and the United States by ISTE Ltd and John Wiley & Sons, Inc.

ISTE Ltd
27-37 St George's Road
London SW19 4EU
UK

www.iste.co.uk

John Wiley & Sons, Inc.
111 River Street
Hoboken, NJ 07030
USA

www.wiley.com

Library of Congress Control Number: 2024949966

British Library Cataloguing-in-Publication Data
A CIP record for this book is available from the British Library
ISBN 978-1-78945-172-6

ERC code:
PE10 Earth System Science
PE10_7 Physics of earth's interior, seismology, geodynamics

Contents

Foreword

Yanick RICARD
LGL TPE, CNRS, University de Lyon, ENS de Lyon, University Lyon 1, France

Silicate mantle makes up most of the Earth's volume (83%) and mass (66%). Its immense thermal inertia controls the planet's slow cooling and has maintained plate tectonics for at least 2 billion years, and perhaps even since the surface solidified a few hundred million years after the Earth's formation 4.54 billion years ago. The causes of plate tectonics, which have not been observed on any of the planets and satellites of the Solar System, remain largely mysterious.

The average properties of the Earth and its mantle have been discussed since the conception of Newtonian mechanics (Cavendish 1731–1810; and Laplace 1749–1827). However, it was not until the beginning of the 20th century that the founders of seismology (Oldham 1858–1936; Gutenberg 1889–1960; and Lehmann 1888–1993) proposed a radial model of its properties. The interest shown by physicists in studying the Earth's interior was not all positive. Determination of the solid character of the mantle (where shear waves can propagate), combined with a lack of knowledge of the phenomena of convection and radioactivity (only discovered at the very end of the 19th century by Becquerel (1852–1903)), was opposed to the concept of "continental drift" and the very long geological timescales predicted by geographers and geologists (on the order of 1 billion years). It was not until the 1940s that the concepts of convective fluid mechanics (Bénard 1874–1939; and Rayleigh 1842–1919) were applied to the mantle with the work of Hales (1911–2006), Pekeris (1908–1993), Hess (1906–1969), or

Runcorn (1922–1995), and isotope geochemistry was able to date geological events (Holmes 1890–1965) and confirm the very long history of our planet. Finally, it was not until the late 1960s that bathymetric and magnetic surveys of the ocean floor made it possible to establish plate tectonics, with the work of Vine and Matthews, Wilson, Morgan, McKenzie and Le Pichon. So, it was only some 60 years ago that mantle was no longer seen as a static object, but as a slowly flowing fluid carrying tectonic plates along its surface.

Although the role of convection in mantle dynamics was accepted and illustrated by various numerical convection models, direct comparison with observations was difficult as long as the three-dimensional structure of the mantle was unknown. In the 1980s, seismologists began to visualize mantle heterogeneities in three dimensions. At the same time, the determination of the Earth's gravity field by studying the trajectories of artificial satellites made it possible to map the gravity field induced by density anomalies that drive convection. Gravity anomalies linked convective fluid mechanics and seismological observations of the mantle. Finally, the ability of numerical simulations to propose models of geological evolution has made it possible to integrate the temporal dimension of observations; paleotectonics and sedimentology, which document the Earth's horizontal and vertical movements, provide clues and constraints for mantle dynamics models. These comparisons between tomography, geodynamics and geology are still a very active field. Also in the 1980s, the refinement of mass spectrometry techniques led to the classification of mantle material sources into a large number of "reservoirs" with subtly differing isotopic or elemental compositions. While geochemistry was so precise in determining many reservoirs, it had no way of constraining their geometry. This view of reservoir geochemistry has lost its relevance as tomographic images and convection models have improved. Although a certain resistance to vertical movements in the mantle is observed at the transition between the upper and lower mantle or, for other authors, more diffusely between 600 and 1,200 km depth, the mantle cannot be described by reservoirs with more or less impermeable boundaries. However, these geochemical studies have left us with two major ideas: indications of recycling of materials that have been on the surface are observed in all magmatic sources, and basalts of hotspot volcanoes are linked to dynamics and sources that differ from those of ridge basalts.

Over the past 20 years, mantle research has both continuously improved its methods and succeeded in approaching its subject with new eyes. In this improvement of methods and tools, the effects of gigantic advances in computer technology are of course to be noted, enabling the construction of models based on millions of data sets, reconstructing the seismic signal over a wide range of frequencies. Computer progress has also democratized convection codes, and convection simulations have been able to take into account many of the mantle's specific features: spherical geometry, high convection vigor, complexity of the rheological law, self-consistent generation of tectonic plates, etc. The ab initio calculation (a chemical model based on quantum mechanics) has also become a new method for exploring mineralogy. Technological advances have also made it possible to study mineralogy in much greater detail at very high temperatures and pressures, and rapid phenomena (phase change, seismic attenuation, fluid percolation, etc.) can now be tracked live on a crystal scale.

New themes have emerged with their own new issues: the differentiation of the Earth during accretion, the nature of the deep mantle, the role of magma extraction in understanding the geochemical signal and the premises of extrasolar geodynamics. In these different fields, seismologists, geochemists, mineralogists, astrophysicists and geodynamicists have had to cross-fertilize their ideas and concepts, and these four lines of research have mutually hybridized.

Geochemistry, with its ever-increasing measurement precision, has revealed very primitive fractionations of the Earth's mantle, totally challenging the idea of a homogeneous primordial mantle. The mantle is neither made up of homogeneous reservoirs sampled by volcanoes or ridges, nor is it the result of the mixing of an oceanic lithosphere extracted from a homogeneous mantle. Experimental mineralogy suggests that core formation took place in a largely molten primitive Earth, where mantle and core equilibrated at depths corresponding to those of the middle mantle. Finally, the observation of thin lenses of fused silicates on the surface of the core, associated with two large zones of the mantle where seismological properties seem to indicate compositional variations, has also changed our view of the mantle. How was the mantle formed? What was its state after core separation? Are the large abyssal provinces of anomalous properties and molten lenses fossils of mantle crystallization? The mantle's early history may hold the key to its present state.

Advances in experimental and ab initio mineralogy, which have clarified the equilibrium conditions of core formation, have also changed our understanding of the deep mantle. While silicate phase changes seemed to take place in the transition zone (between 400 and 700 km depth), and the so-called bridgmanite phase was considered the ultimate and most compact phase of silicates, a new phase has been proposed. This transformation takes place at the same depth as another change, that of the iron spin state, which affects the oxide-perovskite equilibrium. All these changes take place in the deep mantle at pressures close to the core boundary, where the temperature is close to the silicate melting temperature. This makes the zone above the core extremely complex. Of course, this problem of the present-day mantle is closely related to that of the primitive mantle and its evolution since its formation.

The description of the mantle in terms of individualized reservoirs has lost its relevance because magma sources have ceased to be seen as homogeneous environments. Observations at different scales, as well as modeling of convection-driven heterogeneities, suggest an extremely heterogeneous mantle, a "marbled mantle", where chemical and isotopic variability are expressed right down to the grain scale. Thus, the isotopic signal observed at the surface is no longer directly that of a homogeneous source, but the product of a partial melting dynamic whose modeling must take into account the chemical complexity and multi-phase dynamics of magma extraction.

Finally, there is no real scientific understanding when only one object is explained. Understanding the evolution of the Earth alone does not guarantee that geological mechanisms have been understood. Comparative planetology, comparing the Earth with other planets and solid satellites, comparing these with ice and ocean satellites, and finally comparing the planets of our Solar System with those of extraterrestrial planets, opens up new scientific perspectives and will, among other things, have to resolve a major question: What are the conditions that lead to plate tectonics?

1

Models of Mantle Dynamics

Gaël CHOBLET

LPG, CNRS, Nantes Université, France

This chapter provides an overview of mantle dynamics models, from the very concept of movement within the Earth's interior, through the physical formalism introduced to describe it (thermal convection), then the numerical methods used to solve this mathematical problem, and finally the results of such models through a marked evolution in their nature. This chapter does not include a discussion of the data (geophysical, geochemical) that testify to the nature of the mantle, even though the development of models is entirely indebted to these observations.

1.1. Toward the first models of the Earth's mantle

1.1.1. *Prior to the models*

In physics, or more precisely, in mechanics, dynamics is a discipline that groups together the theories that link the motion of a body to the influence of the forces applied to it. As far as the deep dynamics of the Earth's mantle and the planet's interior in general are concerned, it is difficult to trace its genealogy beyond the beginning of the 20th century. However, questions about the age of the Earth, to which the dynamics of the mantle cannot be summed up but which condition it, can be seen as a prehistory.

Structure and Dynamics of the Earth's Interior 1,
coordinated by Julien MONTEUX. © ISTE Ltd 2025.

Without necessarily aiming to emancipate itself from myths and religious narratives, the scientific approach has attempted to include all objects in the universe in its explanatory scheme (hence the name of the latter). Isaac Newton, illustrious craftsman of the universal laws of physics, in order to preserve the model of the Bible, noted that the duration of the first three days was arbitrary, since God only created the stars on the fourth. The scientific approach, however, never ceased to point out the underestimation of the Earth's age by this religious text. While the first stratigraphers to study the deposition of material on the Earth's surface and whose discipline supplies temporal information were initially reluctant to provide absolute dating, preferring no doubt, with good reason, to provide constraints of anteriority or simultaneity, other efforts took up this debate. One example is the (unimplemented) idea of Halley (1656–1742), who saw in the salinity of the oceans, inherited in his view from the flow of rivers, a means of dating. More decisively, there were the studies of the Earth's cooling: Buffon's calorimetric experiments (around 1770) and, finally, Kelvin's theory (in 1862 and then 1897) based on the heat equation that had just been proposed by Fourier.

Assuming, as Buffon did, that the Earth is initially a ball of molten rock (with a temperature of a few thousand degrees) and that its surface quickly reaches a temperature close to that observed today, Kelvin's model proposes the propagation of a cold front by conduction. By simply estimating the thermal conductivity constant of rocks, which Kelvin foresaw might depend on temperature, Fourier's law is sufficient to propose a cooling model in which time and depth are the only variables. We can then compare the variation in temperature at depth, known from the mines (the geothermal gradient, about 3C per 100 m) to identify the time span that led to this evolution. Kelvin's calculation yields an age of several tens of millions of years, well above the Biblical value.

The contribution of geologists to this debate is ambiguous, as their models long struggled in the 19th century to oppose catastrophists and uniformitarians. The latter nevertheless struggled to agree on the estimate of the great physicist Lord Kelvin, which they still found too brief. Similarly, Charles Darwin felt that it would take longer for the evolution of life on Earth to reach the state he observed. It is worth noting that Kelvin was the first to point out that his calculation assumes that the Earth is rigid – much more so than glass, or even steel.

Today, we know (we think) that it is in this supposedly zero movement (the first term in the discipline "dynamics") that the problem lies. Radioactivity, discovered at the beginning of the 20th century, is often mentioned for the final turning point in the age of the Earth, and the heat source it induces is sometimes counted as the major revolution. Rather, it was the dating of rocks that it enabled that was decisive. Kelvin's main error lay in not considering that thermal convection, as yet unformalized, could provide the Earth's mantle with a more efficient means of removing heat, as his assistant Perry had suggested. In a framework where cooling takes place more and more slowly (this is the case for both the conductive and convective models), the time required to reach a given value of flux as measured by the geothermal gradient is longer for more efficient heat transfer. The Earth's age is around 4.5 billion years, and mantle dynamics are the key to this story.

1.1.2. *Continental drift, thermal convection and plate tectonics*

The establishment of the theory of thermal convection applied to the Earth's mantle stems first from the idea of continental drift decisively advanced by Wegener (1915), who argued for horizontal displacements of the planet's surface, compiling a remarkable set of arguments. The acceptance of mantle convection is also, of course, indebted to the understanding of the phenomenon of thermal convection for natural fluids, that is, the movement generated beyond a threshold by the thermal expansion and contraction of a fluid subjected to gravity. The latter was demonstrated in the laboratory in the form of "cellular vortices" in a thin layer of spermaceti (a whale oil that is solid at room temperature) by Bénard (1901)[1] who produced photographs of them (Figure 1.1). Rayleigh (Strutt 1916) then proposed a first mathematical analysis[2] of these meticulous experiments.

Despite the temporal coincidence, the conjunction between the observation of movements on the Earth's surface and the theory of convection in a liquid came up against the concomitant advances in seismology. These demonstrated the great elastic rigidity of the mantle[3] to the point where viscous flow seemed highly improbable even in the eyes of Harrold Jeffreys, one of the most brilliant geophysicists of the time, who had contributed beyond his work in seismology to a better understanding of the phenomenon

1 It has since been established that surface tension between the air and the fluid contributed as much to these experiments as internal variations in density or buoyancy.

2 It is essentially the transient appearance of convection that this first theory describes.

3 With a thickness of 3,000 km above a liquid core, which itself encloses a solid seed (Lehmann 1936).

of thermal convection. The fact that Wegener's proposed driver for continental drift (the Earth's rotation) had proved inaccurate no doubt added to the reticence against it, but less rational objections, listed in detail for the American Earth science community (Oreskes 1999), must also be considered.

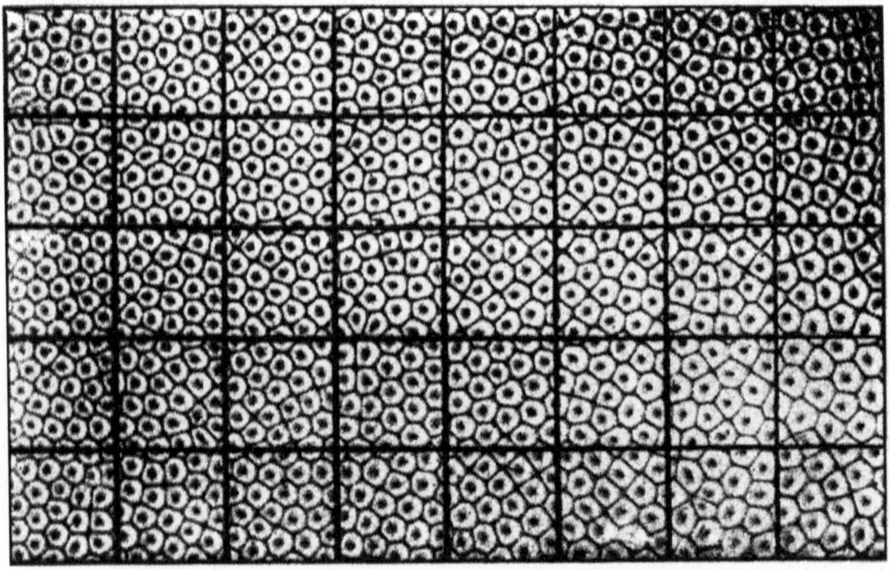

Figure 1.1. *Photograph of convection cells (or cellular eddies) observed by Henri Bénard (Bénard and Avsec 1938)*

Early estimates of mantle viscosity (largely by Jeffreys), although imprecise, nevertheless enabled Holmes (1931) to demonstrate that solid-state mantle convection was a viable mechanism (Figure 1.2). The evaluation of an effective viscosity (3×10^{20} Pa s)[4], describing the vertical creep that seemed to be able to explain the historical uplift of Scandinavia after the melting of a thick ice cap from the end of the Pleistocene some 20,000 years ago (Haskell 1935), confirmed this hypothesis, which remains the current model for understanding the dynamics of the Earth's mantle. The advent of plate tectonic theory, based primarily on paleomagnetic measurements of the ocean floor, has since provided a spectacular framework for the kinematic understanding of movements at the Earth's surface. The intimate modalities of mantle dynamics, however, remain uncertain, as

4 A value that remains reasonable to this day.

the following pages of this chapter illustrate. The history of these joint developments is well documented in Schubert et al. (2001).

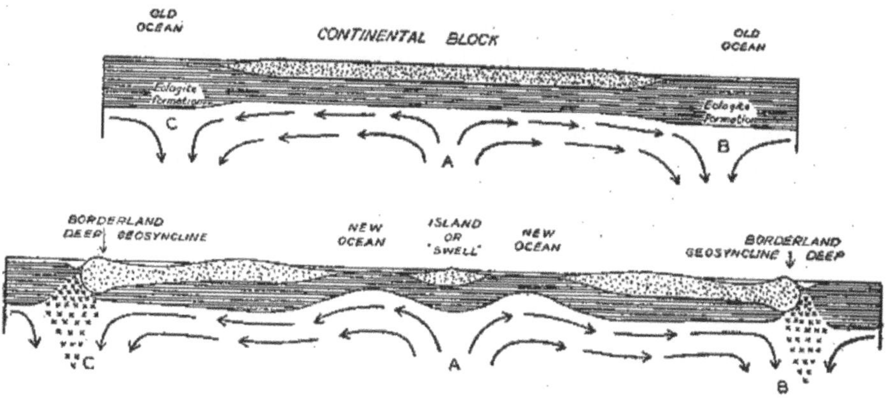

Figure 1.2. *Diagram proposed by Holmes (1931) to explain continental drift through solid-state convection in the Earth's mantle caused by the heating generated by the radioactive decay of the slowest elements, around 20 TW, that is, half the power evacuated by the Earth*

1.2. The physical model: thermal convection

The Earth's mantle is a rocky solid. Like other solids, such as the dough (flour, eggs, water, olive oil) that we put through a rolling mill to make tagliatelle, or the glaciers (water) that remain in certain mountains, it will flow if given enough time: Maxwell's time, formed by the ratio of a material's viscosity η to its stiffness μ (its shear modulus), precisely indicates the duration of mechanical stress beyond which the approximation of liquid behavior becomes reasonable (below which elastic behavior, responsible for earthquakes in the case of the mantle, is observed). Given the age of the Earth, which is much greater than the Maxwell time of the rocks that make up the mantle (around 100-10,000 years), the fluid mechanics approach is well suited to slow movements, although there are still some peculiarities, linked to the very high value of viscosity.

The presentation here is simple and brief. Rich aspects of the physics of mantle convection are described in textbooks (see the very clear, but still reasonably brief, chapter by Ricard (2015)).

1.2.1. *Conservation laws: mass, momentum, energy*

The mathematical relationships describing the phenomenon of thermal convection can be summed up in three conservation equations, which capture its essence. Each of them reflects the fact, in line with human intuition, that in a flow with the velocity field **V** (see Figure 1.3), the only causes of the temporal variation of an extensive mass quantity a within an arbitrary volume Ω are the creation (source) or destruction (sink) of this quantity within this volume, as well as the fluxes of this quantity across the boundary Γ of the volume Ω :

$$\underbrace{\frac{d}{dt}\int_{\Omega}\rho a d\Omega}_{\text{time variation}} = \underbrace{\int_{\Omega}\rho h d\Omega}_{\text{sources/sinks}} - \underbrace{\int_{\Gamma}\rho a\mathbf{V}\cdot\mathbf{d\Gamma}}_{\text{advective flow}} - \underbrace{\int_{\Gamma}\mathbf{F}\cdot\mathbf{d\Gamma}}_{\text{diffusive flow}} \qquad [1.1]$$

ρ [kg m^{-3}] is the density. In the above equation, we have distinguished the part of the flow that is due to the flow (the advective flow) from the part that would occur even for an immobile fluid (the diffusive flow). Negative signs result from the orientation of the surface normal $\mathbf{d\Gamma}$.

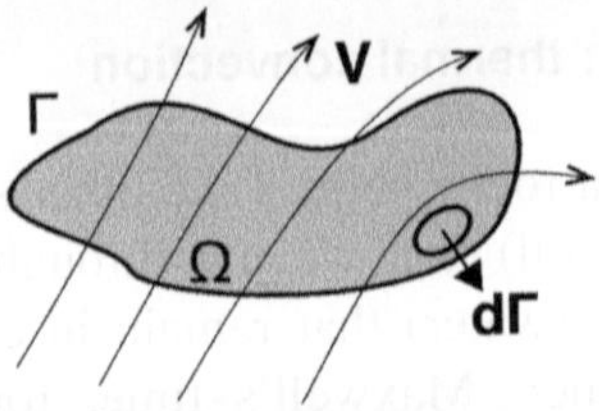

Figure 1.3. *Elementary volume Ω with Γ as boundary, traversed by a flow **V**. $\mathbf{d\Gamma}$ is a vector normal to the surface Γ. Symbols in bold are vectors*

Since no assumptions have been made about the properties of the Ω volume, this global relationship can also be written locally, at each point:

$$\underbrace{\frac{\partial\rho a}{\partial t}}_{\text{time variation}} = \underbrace{\rho h}_{\text{sources/sinks}} - \underbrace{\nabla\cdot\rho a\mathbf{V}}_{\text{advective flow}} - \underbrace{\nabla\cdot\mathbf{F}}_{\text{diffusive flow}} \qquad [1.2]$$

where the divergence theorem is used for both flux terms: the integral of the flux through the closed surface Γ is equal to the volume integral of its divergence on Ω.

This generic local conservation relation can then be applied to the three quantities that describe thermal convection: mass, momentum and energy. In the case of mass, ρa is then formally identified with ρ and in the absence of sources/sinks, and "diffusive" mass flow (an oxymoron):

$$\frac{\partial \rho}{\partial t} = -\ \boldsymbol{\nabla} \cdot \rho \mathbf{V} \qquad [1.3]$$

This relation, also called the continuity equation, is simplified in the case of an incompressible fluid (i.e. for $\rho(\mathbf{x}, t)$ uniform and constant): $\boldsymbol{\nabla} \cdot \mathbf{V} = 0$.

The conservation of momentum ($\rho \mathbf{V}$) is Newton's second law:

$$\underbrace{\rho \frac{\partial \mathbf{V}}{\partial t}}_{\text{time variation}} = \underbrace{\mathbf{f}}_{\text{sources/sinks}} - \underbrace{\rho \mathbf{V} \cdot \boldsymbol{\nabla}\, \mathbf{V}}_{\text{advective flow}} + \underbrace{\boldsymbol{\nabla} \cdot \boldsymbol{\sigma}}_{\text{diffusive flow}} \qquad [1.4]$$

where the continuity equation (equation [1.3]) has been used. We've juxtaposed here the same terms that explained the generic conservation equation, but to get closer to Newton's more familiar principles, we'll note that $\mathbf{f}$ corresponds to volume forces, while $\boldsymbol{\sigma}$ is the stress tensor. The product $\mathbf{VV}$ is also a second-order tensor.

The conservation of total energy (sum of internal energy and kinetic energy) is naturally written in the form presented in equation [1.2]), with the source terms including the work of volume forces and stresses. This equation is most often considered in terms of temperature:

$$\underbrace{\rho C_p \frac{\partial T}{\partial t}}_{\text{time variation}} = \underbrace{\rho H + \boldsymbol{\tau} : \boldsymbol{\nabla} \mathbf{V}}_{\text{sources/sinks}} - \underbrace{\rho C_p \mathbf{V} \cdot \boldsymbol{\nabla}\, T}_{\text{advective flow}} - \underbrace{\boldsymbol{\nabla} \cdot \mathbf{q}}_{\text{diffusive flow}} \qquad [1.5]$$

where an incompressible fluid is again assumed. The $\boldsymbol{\tau}$ tensor of deviatoric stresses (i.e. velocity field-dependent stresses) was introduced to distinguish it from isotropic stresses (i.e. thermodynamic pressure p) $\boldsymbol{\sigma} = -p\mathbf{I} + \boldsymbol{\tau}$ with $\mathbf{I}$ the identity tensor. C_p [J K^{-1} kg^{-1}] is the heat capacity. As we have already seen when discussing the Holmes (1931) model, internal sources of heat ρH related to the radioactive decay of certain isotopes (^{235}U, ^{238}U, ^{232}Th, ^{40}U) play a predominant role. The second source term $\boldsymbol{\tau} : \boldsymbol{\nabla} \mathbf{V}$ corresponds to viscous dissipation, i.e. the conversion of the flow's mechanical energy into heat.

1.2.2. *Constitutive relationships: Fourier, rheology and equation of state*

In order to solve the problem of thermal convection, the three conservation equations [1.3]–[1.5] are complemented by relationships between physical quantities specific to the material, known as constitutive relationships, which describe its response to an external field. These can be deduced from empirical observations or established on the basis of first principles. Potentially complex and often poorly constrained in the case of the Earth's mantle, approximations are often used, as below, for example, by assuming an isotropic response (neglecting the direction dependence of a parameter). In other words, a scalar parameter is introduced when a tensor would better describe the response of the crystals that make up mantle rocks, for example. Similarly, the assumption of a linear response of the material is often proposed.

Fourier's law describes the conductive or diffusive flux of heat, $\boldsymbol{q}$. The canonical form common to many constitutive relations, where the flux is proportional to a gradient,[5] is introduced:

$$\boldsymbol{q} = -k\boldsymbol{\nabla}T \quad [1.6]$$

where the thermal conductivity k (W m^{-1} K^{-1}) is necessarily positive by virtue of the second principle of thermodynamics[6]. If this conductivity is uniform, the diffusive term in equation [1.5] becomes $-\boldsymbol{\nabla}\cdot\boldsymbol{q} = k\nabla^2 T$, with $\nabla^2 = \boldsymbol{\nabla}\cdot\boldsymbol{\nabla}$, the scalar Laplacian. We can then introduce the thermal diffusivity coefficient $\kappa = k/\rho C_p$ [m^2 s^{-1}].

The special role of rheology, that is, the link between the stress tensor and the flow that corresponds to the second constitutive relation, is essential for convection in the Earth's mantle. It will be discussed in more detail in section 1.4.3. It will be shown that a relation of the type:

$$\boldsymbol{\tau} = \eta\left(\boldsymbol{\nabla V} + \boldsymbol{\nabla V}^t\right) = \eta\dot{\boldsymbol{\epsilon}} \quad [1.7]$$

is adapted to the incompressible case. $\dot{\boldsymbol{\epsilon}}$ is the strain rate tensor[7]. Again, the dynamic viscosity coefficient η (Pa s) is necessarily positive and if it is uniform, the stress term in equation [1.4] becomes $\boldsymbol{\nabla}\cdot\boldsymbol{\sigma} = -\boldsymbol{\nabla}p + \eta\boldsymbol{\nabla}^2\boldsymbol{V}$.

5 Denoting zero flux in the case of a uniform field (in this case, temperature).

6 Every source of entropy must be positive.

7 Null for solid rotation or translation.

∇^2 is the vector Laplacian. We can also introduce the kinematic viscosity coefficient $\nu = \eta/\rho$ (m^2 s^{-1}), which could also be called momentum diffusivity, given its formal proximity to thermal diffusivity in the heat equation, whose unit it shares.

Finally, at the heart of the principle of thermal convection, the equation of state links pressure, temperature and density. This seems particularly complex for the materials that make up the Earth's mantle, and for simplicity's sake, a first-order expansion is often employed:

$$\rho = \rho_0 \left[1 - \alpha(T - T_0) + \frac{p - p_0}{K_T} \right] \qquad [1.8]$$

where $\alpha = -1/\rho(\partial\rho/\partial T)_P$ (K^{-1}), $K_T = \rho(\partial P/\partial\rho)_T$ (Pa) are the coefficients of thermal expansion and isothermal bulk modulus, respectively, and T_0, p_0 are the reference conditions. In the case of the Earth's mantle, the second term (pressure effect) causes much greater variations than the first (temperature effect). Nevertheless, these variations proceed radially, in the direction of gravity, and are often neglected for this reason: thermal convection then links the flow to variations in density with temperature alone. The role of composition can be added (see section 1.4.4).

1.2.3. *Boundary and initial conditions*

The three convection equations (equations [1.9]–[1.11]) are partial differential equations (one of which is vectorial) involving three unknowns (one of which is a vector): velocity, pressure and temperature fields. Boundary conditions for the flow and temperature fields are added to these equations. Boundary conditions for flow generally correspond to prescribing the velocity field or the stress field at the boundaries of the domain. Rigid conditions (zero velocity) or free-slip conditions (zero normal velocity, zero tangential stress) are the simplest and most common, even if the actual boundaries are not impermeable. Similarly, temperature or heat flow across the boundary are conditions often considered. In addition to these examples, which can be classified as either Dirichlet or Neumann conditions, a weighted combination of the two (or Robin's condition) is sometimes employed.

Some problems may require boundary conditions that introduce greater realism. This is the case for a free surface[8] if the topography generated by the flow is the key to a problem (rather on a smaller spatial scale than the globality of the mantle). Similarly, in the case of mineralogical phase changes occurring in the transition zone of the Earth's mantle (see section 1.4.5) or a solid/liquid phase change (if a magma ocean is being described, for example), more general interface conditions need to be introduced, characterizing the continuity of certain quantities conserved on either side of the interface. For example, heat flow may be discontinuous, with a jump corresponding to the heat of reaction or phase change.

Finally, an initial condition is required in practice for the temperature field, even if we are only interested in a final equilibrium of the problem, on which it may have only a negligible effect.

1.2.4. *Dimensioning, dimensional analysis and dimensionless numbers*

An essential concept in fluid mechanics is the consideration of similarity laws, that is, the identification of dimensionless numbers (without "unit" or, more precisely, whose SI unit is the number 1) that reduce the number of parameters in the problem. In addition to the clarification that follows, this approach also enables us to study the real phenomenon in experiments on a scale accessible in the laboratory. Taking into account the constitutive relations (equations [1.6]–[1.8]), the dimensionless versions of the conservation equations [1.3]–[1.5] are:

$$v^i_{,i} = 0 \qquad [1.9]$$

$$\frac{1}{\mathrm{Pr}}(v^j_{,t} + v^i(v^j_{,i})) = -p'_{,j} + v^j_{,ii} + \mathrm{Ra}\theta\gamma_j \qquad [1.10]$$

$$\theta_{,t} + v^i\theta_{,i} = \theta_{,ii} + \mathrm{h} \qquad [1.11]$$

The subscripts i (and j) refer to the ith (jth) component of a vector or to the partial derivative along the ith (jth) direction of space, and their repetition means a sum: so $v^i_{,i} = \frac{\partial v_x}{\partial x} + \frac{\partial v_y}{\partial y} + \frac{\partial v_z}{\partial z}$. The index t corresponds to time. The unit vector γ_j is non-zero only in the direction of gravity. All variables are

8 The surface is deformable and stresses, both normal and tangential, cancel out on it.

dimensionless. Thus, the temperature θ is $\frac{T-T_0}{\Delta T}$ where ΔT is the characteristic scale based on the temperature jump across the layer. The velocity v is $\frac{Vd}{\kappa}$ where d is the characteristic length (thickness of the fluid layer) and d^2/κ is the characteristic duration (of thermal diffusion). The pressure p' corresponds only to the dynamic part, the static part balancing the weight. In this system of equations, density is assumed constant in all terms in which it is involved, except in the buoyancy term $\mathrm{Ra}\theta\gamma_j$, which reflects the temperature dependence that drives convection: this is the Boussinesq approximation. This approximation means that the dissipation term (one of the heat sources in equation [1.5]) is neglected in equation [1.11].

Note that the characteristic scales appear only in the three dimensionless numbers Pr, Ra and h:

$$\mathrm{Pr} = \frac{\nu}{\kappa} \qquad [1.12]$$

$$\mathrm{Ra} = \frac{\alpha g \Delta T d^3}{\kappa \nu} \qquad [1.13]$$

$$\mathrm{h} = \frac{hd^2}{k\Delta T} \qquad [1.14]$$

The Prandtl number (Pr) therefore expresses the ratio between two diffusivities and depends solely on the material in question. The rocks that make up the Earth's mantle are relatively heat-insulating ($\kappa \simeq 10^{-6}$ m^2 s^{-1}); conversely, the extreme magnitude of viscosity ($\nu \simeq 10^{17}$ m^2 s^{-1}) reflects an almost instantaneous diffusion of momentum[9] and means that, in practice, Pr can be considered infinite. Conservation of momentum (equation [1.10]) is therefore reduced to its right-hand term: $0 = -p'_{,j} + v^j_{,ii} + \mathrm{Ra}\theta\gamma_j$, which fundamentally simplifies its treatment[10]. On the other hand, a great deal of the complexity that characterizes flow is erased in this simplified form, which assumes that viscosity is uniform. We will see later how special the rheology of mantle fluid is (section 1.4.3).

The Rayleigh number (Ra) is the fundamental number of thermal convection. It measures the competition between the driving force (buoyancy,

9 Absence of inertia: the kinetic energy of a tectonic plate is less than that of an average vehicle on a Vendée road.

10 This inertial term can give rise to nonlinear instabilities leading to turbulence, a particularly complex physical problem.

to which we can associate the characteristic time $\tau_b \sim \sqrt{d/\alpha g \Delta T})$ and the processes that oppose it (the diffusion of heat and momentum of characteristic times $\tau_c \sim d^2/\kappa$ and $\tau_v \sim d^2/\nu$, respectively). This dimensional analysis reflects the fundamental aspect of thermal convection as a threshold phenomenon: in order for instabilities specific to convection to develop, the ratio $(\tau_v \tau_c)/\tau_b^2$ must be sufficiently large, that is, the Rayleigh number must exceed a critical value, approximately equal to $\text{Ra}_c \simeq 1{,}000$. Note that the Rayleigh number involves parameters specific to the fluid (α, κ, ν), but also other quantities, characteristic of the set-up (g, ΔT, l). For the Earth's mantle, it is about $10^7 - 10^8$, guaranteeing the existence of convection.

1.2.5. *From linear stability analysis to turbulent convection*

Solving the problem of thermal convection, expressed in the set of equations [1.9–1.13], includes several approaches that can be illuminatingly associated with different stages of the phenomenon, depending on the value of the single parameter that is the Rayleigh number.

Rayleigh's pioneering theoretical work was aimed at understanding the appearance of convective movements at the threshold quantified by the critical value Ra_c. A linear stability analysis is produced, which, considering small perturbations of a conductive equilibrium state that is always a solution of the problem ($\mathbf{v} = \mathbf{0}$), allows the study of a linear approximation of the system of equations: below the value Ra_c, perturbations decrease whatever the spatial wavelength, and equilibrium is therefore stable.

Just above the Ra_c threshold value, rolls appear, invariant in a horizontal direction, appear, in line with the predictions of weakly nonlinear models developed beyond the linearized problem. As the value of Ra continues to increase, the nature of the dynamics gains in complexity. Three-dimensional stationary forms are observed (hexagons) that break the vertical symmetry of the solution[11], then time-dependent periodic, then strongly unsteady flows with intermittent plumes (see Figures 1.4 and 1.5).

11 These shapes are also favored when an ingredient in the set-up makes it asymmetrical. This is the case in the laboratory experiments shown in Figure 1.5, where fluid viscosity depends on temperature.

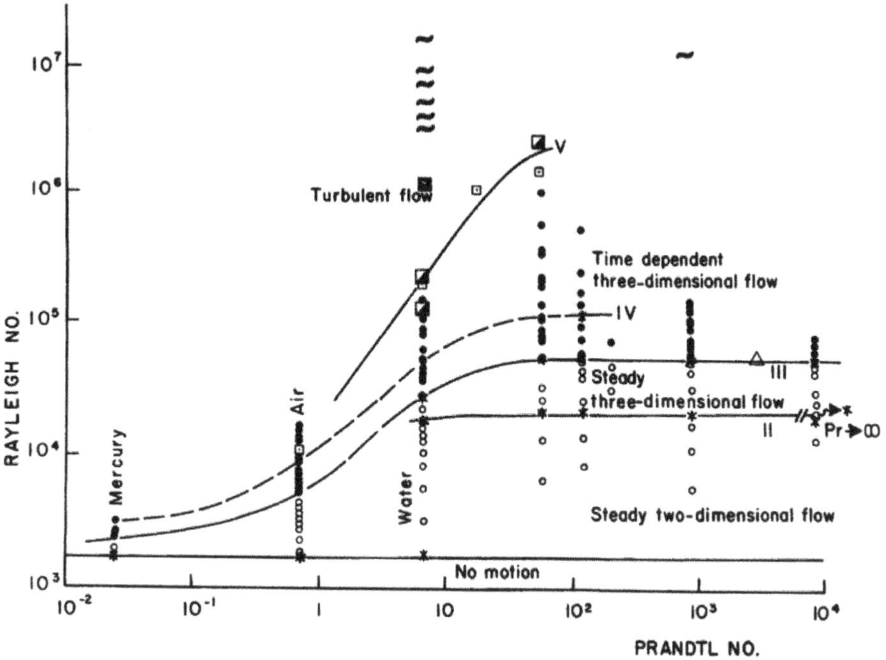

Figure 1.4. *Flow diagram for convection patterns*

COMMENT ON FIGURE 1.4.– *The x-axis shows the Prandtl number – the values associated with certain fluids are indicated, with the Earth's mantle at the far right of the diagram; the y-axis shows the Rayleigh number, whose increase leads to convection crossing the threshold into 2D, then 3D, then unsteady convection which becomes turbulent (figure taken from Krishnamurti and Howard (1981), consult for the meaning of the symbols).*

Access to these solutions is only possible through numerical simulation or laboratory experiments. There is, however, another approach that should be mentioned, corresponding to the analytical resolution of these problems. A very clear account of this kind of approach to geodynamics is given by Ribe (2015): it can include the resolution of simplified problems, which allow a deep understanding that is often partly obscured by the numerical approach. Comparing the results of a simulation tool with the analytical solutions of a Stokes flow (equations [1.9] and [1.10] without coupling with conservation of energy) is also a way of testing how well it works[12]. In addition, the formulation of a dimensional analysis as was done for thermal convection (section 1.2.4) is often required for the analysis of the physical problem. Finally, beyond dimensional analysis, one of the modalities of the analytical approach, scaling analysis, enables a deeper understanding of simulation results. In the case of thermal convection, such an exercise is based on the estimation of the orders of magnitude involved in the equations, and leads, for example, to the establishment of the scaling law relating the Nusselt number, an adimensional measure of the mean heat flux q across a convection boundary:

$$\mathrm{Nu} = \frac{qd}{k\Delta T} \qquad [1.15]$$

to the Rayleigh number (and the Prandtl number). Beyond the critical value of Ra, the heat transfer efficiency obtained for solutions of equations [1.9–1.11] increases like a power law:

$$\mathrm{Nu} = A\mathrm{Ra}^{\beta} \qquad [1.16]$$

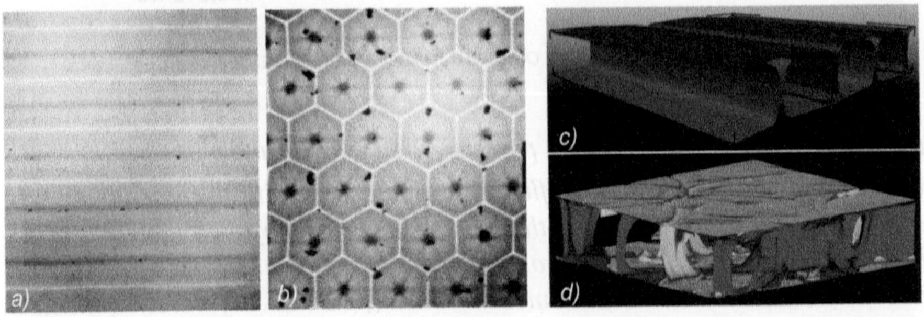

Figure 1.5. *Different forms of convection. (a, b) Shadowgraphs of laboratory experiments (top view, White (1988)): stationary rolls (2D) and hexagons (3D). (c, d) Numerical simulations by S. Labrosse: stationary rolls (2D), unsteady plumes (3D). Two isothermal surfaces are shown, one hot (red) and the other cold (blue). For a color version of this figure, see www.iste.co.uk/monteux/earth1.zip*

For high values of the Rayleigh number, the thickness δ of convective structures (plumes and the boundary layers from which they originate, see

12 See, for example, the analytical solutions derived by Kramer et al. (2021) for a spherical shell.

below) becomes thinner and independent of the total thickness d. Consequently, so does the heat flux $q \simeq k\Delta T/\delta$ and, as equations [1.15] and [1.16] entail:

$$q = A\frac{k\Delta T}{d}\left(\frac{\alpha g \Delta T d^3}{\kappa\nu}\right)^\beta \quad [1.17]$$

we obtain that $\beta \simeq 1/3$.

Figure 1.6 proposes a simple phenomenological model for a convective layer, that of the cell (see, for example, (Turcotte and Oxburgh 1967)). The most important structures are the two boundary layers, with a thickness of δ in the vicinity of the impermeable horizontal boundaries: they can be defined as the regions where thermal conduction is the predominant contribution to heat transfer, and it is here that vertical temperature variations originate[13]. Between these two layers, vertical flow is developed and the vertical temperature gradient is zero (also known as isothermal flow). Advective flow predominates. Note that in this case, where all the heat is supplied at the base of the layer (where $T_1 = T_0 + \Delta T$ is prescribed, T_0 being the surface temperature), the heat flux passing through any horizontal plane is always equal to q, the surface value: the sum of the two contributions (advective and diffusive) is invariant. Thermal plumes have a horizontal dimension comparable to the thickness of the boundary layer from which they emanate. Each carries half the heat in a symmetrical set-up. In the case of an infinite Prandtl number and for an isoviscous fluid, this dimension is smaller than the characteristic scale for lateral variations in vertical velocity: plumes are not flow conduits.

This model enables us to produce a scaling analysis that also allows us to establish equation [1.16]: an analytical solution of the flow in the isothermal core is indeed accessible and, once the principles outlined above are introduced, an equilibrium of forces within each plume[14] translates into the scaling law between Nu and Ra.

13 For a symmetrical set-up, each boundary layer accounts for half of the total temperature increase.

14 Between buoyancy, the driving force and viscous constraints.

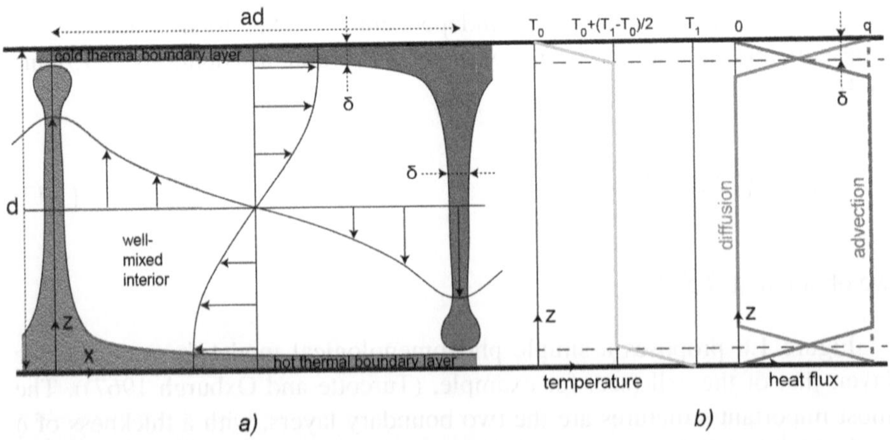

Figure 1.6. *Convection cell model (Rayleigh Bénard case for an isoviscous fluid with free-slip). (a) Schematic diagram with thermal structure, flow and characteristic lengths d and δ (a is the aspect ratio). (b) Vertical profiles of temperature and heat flux (advective and diffusive, the sum is shown by the purple dashed line). For a color version of this figure, see www.iste.co.uk/monteux/earth1.zip*

1.3. "Solving" partial differential equations

For simplicity's sake, we will note the problem of equations [1.9]–[1.11]:

$$R(u) = f \qquad [1.18]$$

where the operator R gathers the various partial derivative terms, u is the solution sought at any point of a domain Ω which can be 2D or 3D, Cartesian or spherical and f is the second member. We add to this problem the boundary conditions on the boundary of the Ω domain.

In practice, solving this problem means using numerical computation to find an acceptable approximation to the solution (without worrying about its uniqueness, or even its existence!): it is this kind of numerical model that is discussed in this section. Before describing these briefly[15], let us finally note that the nature of the equations makes explicit treatment of the time step[16].

15 More details are available in several textbooks (see, for example, (Ismail-Zadeh and Tackley 2010)).

16 Although t is a continuous variable, numerical solutions often offer only a discrete approximation with a δt interval (the time step), most often variable, between two consecutive instants. The principle is explained in more detail in section 1.3.2.

Since only the temperature field involves a time dependence (equation [1.11]), a practical approach is, at each time step, to solve the flow through equations [1.9] and [1.10][17] and use the velocity field thus evaluated to describe the advection and find out the temperature field at the next instant.

1.3.1. *Spectral methods*

This set of methods was implemented fairly early on to describe convection inside planets, because of its proximity to the analytical solutions that prevailed before humans had sufficiently efficient computers. In practice, it means decomposing the solution of equations onto a base of orthogonal functions.

We have also seen that equations [1.9] and [1.10] involve the Laplacian operator[18] if diffusivities are uniform. Thus, in a spherical shell, spherical harmonic functions, which are precisely the eigenfunctions that solve the Laplace equation in such a geometry, are used for spectral decomposition in azimuthal directions. Significant accuracy can be achieved, measured by the harmonic degree[19] of truncation to which functions are developed, at relatively modest cost. The spatial derivatives that appear in the equations correspond to analytical expressions and, in the case of linear equations, the different harmonic modes are decoupled in the spectral domain and can therefore be treated independently. If products of functions are involved (such as the advective term in energy conservation), recourse to physical space (and meshing, see below) may be required. In the vertical, or radial, dimension, for which boundary values are prescribed, these methods use a variety of approaches (Chebyshev polynomials, or grid methods such as finite differences). The first codes in spherical (3D) geometry were built on this type of method (see, for example, (Machetel et al. 1986)).

The spectral approach loses its elegance in cases where diffusivity coefficients, notably viscosity, vary laterally. Although palliatives are conceivable, they do not appear to be effective beyond a threshold for the

17 The buoyancy (as well as the viscosity if temperature dependence is taken into account, see section 1.4.3) is then known at time t.

18 Conservation of momentum, for example, can even be simplified by introducing a potential, the stream function φ such that $\mathbf{v} = \nabla \times \varphi$, which is the solution of a biharmonic equation involving the square of a Laplacian.

19 Inverse of wavelength.

viscosity gradient, and these methods have become less popular as more realistic rheologies (see section 1.4.3) have been considered. Nevertheless, they are still widely used in geodynamo codes.

1.3.2. *Finite-difference or finite-volume methods*

Finite-difference methods are the oldest, having already been envisaged by Euler in the 18th century: the unknown continuous space and time functions (velocity, pressure and temperature vectors) in the conservation equations are replaced by a collection of discrete values that can be solved numerically. Partial derivatives are then approximated by differences: Figure 1.7(a) shows how a uniform 2D mesh is produced for the plane with spatial increments Δx and Δy. Pressure p, for example, is considered only through point values (yellow disks) on the mesh shown; the component in the x direction of the pressure gradient, evaluated at a point in the middle of the mesh points (red cross), is thus translated as:

$$\frac{\partial p}{\partial x} \simeq \frac{p^{i+1,j} - p^{i,j}}{\Delta x} \qquad [1.19]$$

All the differential operators present in the equations and boundary conditions are therefore replaced by such approximations, usually derived from Taylor expansions. This approach is intuitive and easy to program. The main obstacles are questions of precision (obviously linked to the size of the mesh, but also to the order of the Taylor expansion) and convergence (the difference between the approximate solution and the exact solution must tend toward 0 when the mesh becomes infinitely small), which are generally translated into a stability criterion for the numerical scheme.

The whole problem is translated by matrix equality:

$$Lu = f \qquad [1.20]$$

where L is a matrix equivalent to the operator translating the set of spatial derivatives, u is the solution vector on the mesh considered, grouping together the pressure p and the components of the velocity vector **v** or the stream function φ, depending on the formulation adopted, and f is the right-hand side which includes in particular the buoyancy term ($Ra\theta\gamma_j$ for the velocity-pressure formulation, equation [1.10]). This matrix contains a large number of null values, and the coefficients relating to the discretization scheme are arranged in diagonal bands. This is the approach used in early mantle convection codes (e.g. (Torrance and Turcotte 1971; McKenzie et al. 1974)). The matrix conditioning that accounts for the difficulty of the problem is deteriorated by the introduction of strongly spatially variable coefficients: this is the case when viscosity is highly temperature dependent. In practice, this system can be solved with a direct approach for the 2D case, notably with matrix decomposition methods. When this is not possible, as with the 3D problem whose size (several million unknowns) precludes a direct approach, an iterative method is used – multigrid schemes (see section 1.3.4) are then an almost systematic recourse.

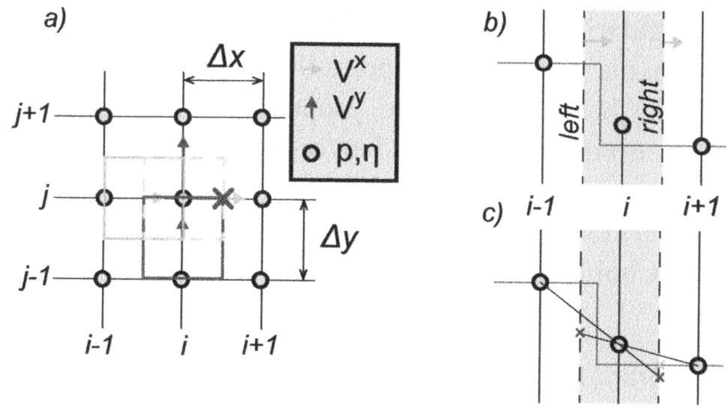

Figure 1.7. *Some aspects of finite-difference and finite-volume methods. For a color version of this figure, see www.iste.co.uk/monteux/earth1.zip*

COMMENT ON FIGURE 1.7.– *(a) Example of a regular 2D mesh. The principle of staggered grid mesh is shown (with variables described in the title block), as are the associated control volumes, represented by colored squares. (b and c) Principle of flux-limiting methods for a 1D problem: (b) fluxes are evaluated through the control volume (shaded and bounded by dashed lines, left and right) corresponding to the values shown on the y-axis, with velocities flowing to the right. (c) Flux problem on the right: a second-order scheme extending the upstream values $i-1$ and i leads to a value smaller than the local minimum (red cross on the right); the flux must be limited (blue cross).*

Finite-volume methods take up part of the finite-difference approach, notably the approximation of differential operators by discrete values on a mesh. However, conservation of a quantity is envisaged as its balance across a control volume (e.g. yellow dotted square in Figure 1.7a): fluxes across volume boundaries are discretized, and source/sink terms are considered inside the volume. In practice, this approach leads to a matrix problem of the same type as equation [1.20]. It is often useful to consider an staggered-grid positioning of the variables, with scalar quantities (pressure, temperature) at the center of the cells (yellow disks) and velocity components at the center of the facet normal to the direction (cyan arrows): the control volume is then different for each conservation equation (yellow, cyan and red squares in Figure 1.7(a)).

In the case of 3D spherical shell geometry, particular care must be taken in choosing the mesh and coordinate system used, since spherical coordinates induce polar singularities: cells bounded by constant longitudes become arbitrarily small in the vicinity of the poles. Difference or finite-volume codes for such geometries have, for example, proposed the Yin-Yang configuration, where spherical coordinates are used over two baseball-like regions, thus avoiding the pole problem (e.g. STAGYY code (Tackley 2008)), or the use of orthogonal coordinates over six identical blocks partitioning the sphere (e.g. OEDIPUS code (Choblet et al. 2007)).

1.3.3. *Finite element methods*

Finite element methods use some of the same ingredients as finite-volume methods, such as the principle of meshing. This method takes advantage of a variational formulation of the problem in equation [1.18], which can be written as follows: find u on the domain Ω such that:

$$\int_\Omega R(u)v = \int_\Omega fv \qquad [1.21]$$

regardless of the function v defined on Ω. A solution of equation [1.18] is necessarily also a solution of equation [1.21], but the opposite is only true for sufficiently smooth functions (for this reason, we speak of a weak formulation). This second problem is then discretized, and only non-zero piecewise polynomial functions on the element under consideration are sought on the discrete subspace associated with the mesh. Finite volume approaches are, in a way, a linear version of finite element methods, but the latter give access to the solution over the entire domain (and not just at the mesh points). However, the solution can be reduced to a matrix problem similar to equation [1.20].

An important advantage of the finite element approach is that it allows the use of meshes with complex geometries, either because the study domain is complex, or – and this is particularly the case for mantle dynamics models – if we wish to use adaptive meshing (see section 1.3.4) to focus effort on certain regions. The CITCOMs code, which solves the problem of thermal convection of a fluid with variable viscosity in a 3D spherical shell (Zhong et al. 2000), and the more recent collaborative code ASPECT (Heister et al. 2017) (see also below) are based on finite element methods.

1.3.4. *Various aspects of contemporary methods*

This section discusses some of the salient aspects of the most recent codes dedicated to the study of mantle dynamics, both finite element and finite volume approaches.

1.3.4.1. *Multigrid methods*

We have seen that, given the cost of random access memory, the matrix problem [1.20] is solved by iterative methods. An effective way of speeding up convergence is to use the multigrid approach, the principle of which is to use coarser meshes to smooth out the longest wavelength error on the solution of the iterative problem. If u_h^n is the approximate value of the solution of the discrete problem u_h^* at step n on the mesh grid indicated by the index $\bullet_h$, the error at this step is $e_h^n = u_h^* - u_h^n$. Unknown by nature, its amplitude is nevertheless accessible through the value of the residual r_h^n :

$$r_h^n = L_h e_h^n = L_h(u_h^* - u_h^n) = f_h - L_h u_h^n \qquad [1.22]$$

We then proceed to restrict operator R of the residue on a coarser grid indicated by the index $\bullet_H$, $r_H^n = \mathcal{R}(r_h^n)$ in order to solve the less costly problem:

$$L_H e_H^n = r_H^n \qquad [1.23]$$

After a few smoothing iterations, e_H^n is extrapolated (or extended $\mathcal{P}$) onto the finer initial grid. This process can be iterated – known as a multigrid cycle – with sufficient frequency, as the $\mathcal{R}$ and $\mathcal{P}$ operations introduce a bias on the resolution of the finer grid. This particularly effective method[20] was first

20 The speed of convergence is proportional to the size of the N matrix ($\mathcal{O}(N)$), whereas that associated with classical iterative methods is generally $\mathcal{O}(N^{3/2})$-$\mathcal{O}(N^2)$.

used on a 3D Cartesian grid and for an isoviscous fluid (with a formulation for the stream function) by Parmentier and Sotin (2000), and subsequently introduced into most codes. In the case of strong viscosity gradients, however, the use of excessively coarse grids, where variations in this coefficient are poorly described, become less effective.

1.3.4.2. *Advection scheme*

A source of difficulty is also associated with the treatment of the advective term in energy conservation (term $v_i\theta_{,i}$, equation [1.11]) on a mesh: numerical oscillations or the introduction of numerical (non-physical) oscillations are known problems of the simplest schemes. Several approaches have been developed to overcome these drawbacks, including flux-limiting methods (see Figure 1.7b) for higher order (high-resolution) methods.

1.3.4.3. *Particle in cell (PIC) method*

When the advection term is included in a flux equation that does not include a diffusive term (this is the case for the transport of compositional heterogeneities, see section 1.4.4), the high-resolution methods just mentioned can remain problematic if they artificially introduce even the slightest diffusion of a field that does not exist. One recourse has been the use of a semi-Lagrangian scheme in which tracers (particles) are introduced into the domain carrying the characteristic value of the transported field. These tracers are advected with great precision by the velocity field known on the mesh, and interpolated to reconstruct the values of the transported field on the grid. In the case of mantle dynamics models, this approach was proposed for a velocity-pressure formulation by Gerya and Yuen (2003) and frequently taken up thereafter.

1.3.4.4. *Parallelism*

The increasingly widespread use of 3D models, often in spherical geometry, and the associated increase in the size of problems in terms of memory space and calculation time have been accompanied by the adaptation of tools for operation on parallel computers[21]. The use of spectral methods, less straightforward to implement in such a setting, loses some of its appeal in this respect, even if the problem of variable viscosity remains the real limitation. Mesh-based methods (finite elements and finite volumes), on the other hand, offer a natural implementation for this type of operation, notably

21 Multiple processes, each with its own memory, can be performed simultaneously.

by decomposing the spatial domain into sub-domains where the same mathematical operations are performed: all that is required is for the various parallel processes to exchange data at the (virtual) boundaries of these sub-domains, which is easily achieved through the use of MPI (message passing interface) libraries. This approach delivers very high computational efficiency as the mesh size increases, involving more powerful computers (allowing a greater number of processes to run in parallel).

1.3.4.5. *Adaptive mesh*

As we have seen, the size of convective structures decreases as the Rayleigh number increases. Thus, in the Earth's mantle, the large-scale spatial movements manifested at the surface by the size of tectonic plates (thousands of kilometers) are also associated with much finer scales, such as hot plumes and subducting plates and the thermal boundary layers from which they originate, which are of the order of a hundred kilometers thick. In addition, the rheological characteristics of rocks mean that deformation is even more localized (section 1.4.3). Compositional heterogeneities (section 1.4.4), although still poorly understood, may also occur on small spatial scales. For this reason, the use of quasi-uniform meshes does not allow us to effectively cover the full range of scales (since solving the finest ones would entail a computationally prohibitive number of cells), and one needs to consider local refinement. Furthermore, while the localization of some of the smallest structures (e.g. thermal boundary layers) is easily conceivable a priori, others emerge in regions that result from system dynamics and vary over time. The solution is therefore to allow dynamic refinement during calculation, in areas where strong gradients (in viscosity or composition) are involved. Adaptive mesh tools of this kind are beginning to appear: the methods that make them possible are, however, complex, particularly in the context of parallel computing, where local increases in mesh size (or decreases in cell size) lead to load considerations for each of the computational processes, which we hope will be equitable, so as to associate a globally comparable computation time with each one.

1.3.4.6. *Development of collaborative tools*

One way of dealing with these growing complexities is to build collaborative tools where the development of new functionalities is made possible by a large number of contributions from researchers using the tool. One example of this is the ASPECT code (Heister et al. 2017), the number of users of which has been growing in recent years.

1.4. From a reductionist to a holistic approach

The first numerical models produced in the 1970s for the dynamics of the Earth's mantle (Torrrance and Turcotte 1971; McKenzie et al. 1974) produced some very encouraging findings. For instance, the orders of magnitude of observables such as tectonic plate velocities, heat flux and even gravity anomalies predicted by these calculations seemed to be in agreement with the estimates made at the time. Since then, the measurements that can be compared with these models have become much richer (suffice it to consider the progress made since then in seismology or geochemistry), and the models have incorporated more ingredients as numerical methods and computers have made them possible. The nature of more recent models has also changed. However, early efforts are focused on the effect of one parameter or another, and circumscribing it, even if it meant drawing only marginal conclusions about its application to the Earth's case, more recent work often includes a large number of more realistic factors to enable more direct comparison with one observation or another. Undeniably, clarity has often been lost in terms of the individual processes at work and their deeper understanding. The context of this volume does not allow us to report on all these works, or even to mention them all. Instead, the following pages suggest a simple path through some of these models to convey how this march toward complexity proceeds, from the exposition of simple equations (equations [1.9]–[1.11]) to the most recent advances.

1.4.1. *Mixed heating*

A first aspect not mentioned so far concerns an essential feature of the Earth's mantle as a convection set-up, namely, the nature of the heating. For the Earth, we now consider that of the heat that crosses its surface (around 45 TW) (see Chapter 5 by Stéphane Labrosse for more details), part (around 10 TW) comes from the heat supplied by the core at a temperature of around 4,000 K at the base of the mantle. However, the greater part comes from additional heating, caused by the radioactive decay of certain abundant isotopes in the rocks[22] and by the average cooling of the mantle, which can be considered as a heat source in its interior.

22 The distribution is highly heterogeneous, with around a quarter of this heating concentrated in the upper part of the lithosphere, which accounts for less than a 20th of the volume.

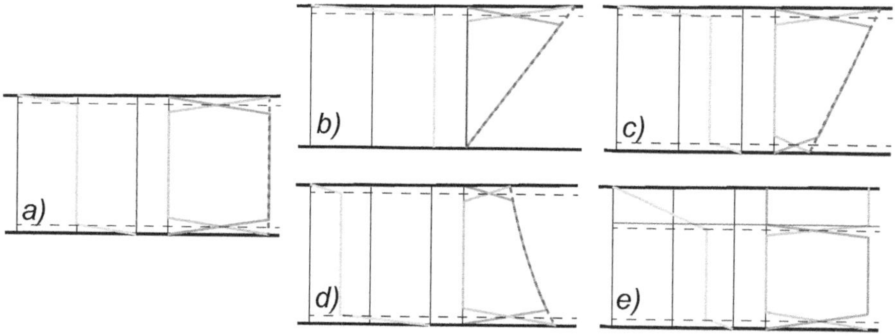

Figure 1.8. *Schematic vertical profiles of temperature (yellow curve) and vertical advective (green) and diffusive (cyan) flux, as well as the sum of the two contributions (purple dashed line). For a color version of this figure, see www.iste.co.uk/monteux/earth1.zip*

COMMENT ON FIGURE 1.8.– *(a) For basal heating, (b) for internal heating, (c) for mixed heating, (d) for a spherical layer and (e) for the conductive lid regime of a cartesian layer (fluid whose viscosity varies strongly with temperature).*

It follows that the simple diagram outlined above (Figures 1.6 and 1.8(a)) needs to be adapted. In the case where heat sources are produced entirely within the fluid layer, the hot thermal boundary layer disappears (Figure 1.8(b)) and only cold instabilities are produced, with the mass-conserving return flow taking the form of a relatively diffuse passive upwelling of hot material. The vertical heat flux, zero at the base of the layer, then increases gradually (linearly if the heating rate is homogeneous, see the sum of the advective and diffusive fluxes in purple in Figure 1.8(b)) as one rises toward the top of the layer. If heating is mixed (basal and internal heating, Figure 1.8(c)), the situation is intermediate. The cold boundary layer exists, but is associated with a temperature jump of less than half the global temperature jump – the symmetry of the problem is broken.

1.4.2. *Sphericity*

The same schematic approach can be applied to the influence of sphericity. Once again, the symmetry between the cold and hot regions (boundary layers and convective instabilities) is broken by the fact that the greater the curvature of the spherical shell, the smaller the area. This curvature can be measured by the ratio of the radii corresponding to each of the spheres that make up the boundary $\zeta = r_b/r_s$, where r_b and r_s are the radii at the base and surface, respectively. In the absence of internal heating, the powers passing through the two surfaces (and each of the inner spheres of intermediate radius) being equal at equilibrium, the flux passing through the surface is ζ^2 times smaller than that prescribed at the base. This has the opposite effect to that of internal heating: the average temperature is reduced because the jump across the hot layer is greater (Figure 1.8(d)). Similarly, the flow is affected by the change in area as the radius varies, and the radial/vertical velocity maximum, corresponding to mid-height in the case of a flat layer (Figure 1.6), is shifted toward the upper interface.

1.4.3. *Rheology*

The two characteristics just described (mixed heating, sphericity) for the Earth's mantle, although leading to major deviations from the canonical model of a flat layer heated at its base for an isoviscous fluid, remain simple effects. The same cannot be said of considerations relating to the rheology suitable for describing the Earth's mantle. While the fact that, in practice, rocks are associated with a very high viscosity is a benefit that has already been mentioned, and helps to simplify the treatment of the problem, the complexity of rheology and a number of aspects that are still poorly understood make it an arduous task.

A general law describing the viscosity of a rock of given composition, a polycrystalline solid, is as follows for a temperature corresponding to a sufficient fraction of its melting temperature:

$$\eta = A\tau_{II}^{1-n} b^{-m} \exp\left[\frac{E + pV}{RT}\right] \qquad [1.24]$$

where τ_{II} (Pa) is the square root of the second stress tensor invariant[23] $\boldsymbol{\tau}$, b (m) is the grain size, E (J mol^{-1}) and V (m^3 mol^{-1}) are the activation energy and volume, respectively, and R (J mol^{-1} K^{-1}) is the perfect gas constant, and n and m are exponents (of stress and grain size).

In the case of a Newtonian fluid, that is, one whose viscosity depends only on its nature, not on the flow, and for which the stress exponent n is 1, the first

23 A scalar representative of their amplitude: $2\tau_{II}^2 = \mathrm{tr}(\boldsymbol{\tau})^2 - \mathrm{tr}(\boldsymbol{\tau}^2)$.

contribution to viscosity variations is temperature via the Arrhenius form (the exponential term), which reveals a thermally activated process. Thus, creep experiments indicate that variations of thousands of degrees in the mantle must cause variations of several orders of magnitude in viscosity – the latter can be divided by 10 if the temperature rises by 100°C. From this point of view, the lithosphere can be identified as the cold thermal boundary layer, which is considerably more viscous, thus accounting for both its mechanical and thermal properties. Models involving such variations show that a particular regime of thermal convection then comes into play (Figure 1.9), where the cold boundary layer becomes sufficiently viscous to no longer participate in the flow (red horizontal line in Figure 1.8e above which advective flow is cancelled out: heat transfer proceeds solely by conduction). Convection is then confined to a warm sub-layer where viscosity varies little (from 1 to 10). Most of the temperature and viscosity variations take place above this sublayer. The effect is to heat up the convective core, albeit in a different way from internal heating (Figure 1.8(c)).

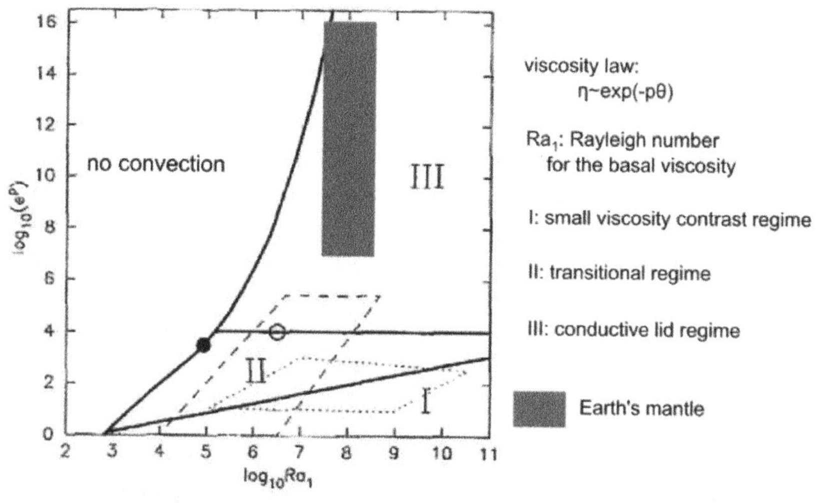

Figure 1.9. *Regime diagram for convection with temperature-dependent viscosity. The x-axis (logarithmic) shows the Rayleigh number. The y-axis shows the viscosity contrast. For a color version of this figure, see www.iste.co.uk/monteux/earth1.zip*

COMMENT ON FIGURE 1.9.– *Below a critical value, which depends on the viscosity contrast, there is no convection. Three convective regimes are expected, separated by bold lines, including the conductive lid regime. Adapted from Solomatov (1995).*

Remarkably, the inclusion in the models (at considerable computational cost, see section 1.3) of this prime ingredient, the strong variations in rock viscosity, has first led to the results being distanced from an essential aspect of mantle dynamics that previous simplistic models managed to reproduce, namely, surface mobility. Although the movement of tectonic plates, where deformation is confined (to first order) to narrow boundaries, is not identical to that of an isoviscous fluid subjected to thermal convection[24], it remains that the formation of a cold, viscous lid on the (immobile) surface, while undoubtedly accounting for the rigidity of the blocks that these tectonic plates constitute, above all leads to a drop in efficiency for the transfer of heat[25] that the subduction of cold material favors in the plate tectonic regime.

A well-known complexity, although still poorly addressed by models, is that rock creep mechanisms do not boil down to the simplest Newtonian case. For the Earth's mantle, let us mention the two main ones: (1) creep by diffusion of crystalline vacancies (or conversely, atoms) and (2) the motion of linear or planar crystalline imperfections, known as dislocations. The former, which reflects a flow of mass from crystal faces in compression to those in extension, is similar to a diffusive process and can be considered as Newtonian behavior ($n = 1$, equation [1.24]). However, the associated viscosity depends on grain size b ($m = 2 - -3$). Conversely, dislocation motion does not depend on b ($m = 0$, although it does induce changes of this size) but on another distance scale, the characteristic dislocation spacing, which in turn depends on the applied stresses ($n \simeq 3$). It is expected that the creep mechanism associated with the lowest viscosity for given conditions will predominate. For example, in the case of olivine, the most abundant magnesium-iron silicate crystal in the Earth's upper mantle, it is likely that dislocation creep is favored in the upper parts, but that diffusion creep takes place at depth. However, a satisfactory treatment of grain size evolution has yet to be established.

One form of convection within planetary mantles that makes this fluid unique is toroidal plate motion: as soon as the field has zero divergence, velocities can be decomposed into two components (poloidal and toroidal). The first corresponds to the sources and sinks for surface velocity divergence,

---

24 Another major discrepancy concerns oceanic ridges which, although the site of surface divergence, correspond to passive upwellings caused by plate spreading rather than hot plumes.
25 We can simply say that the value of the exponent β translating the efficiency of the flow to cool the fluid (equation [1.16]) becomes well below 1/3.

precisely the expression of convective dynamics expressed in the vicinity of ridges (source) or subduction zones (sink); the second typically involves rotation of solid blocks around vertical axes, as seen in the vicinity of transform boundaries on the Earth's surface. This component cannot be produced by convection of a fluid without lateral variations in viscosity, and the amplitudes measured for the Earth (Dumoulin et al. 1998) do not seem accessible to the usual viscous rheologies, even if the stress exponent n had a high value.

Our discussion of mantle rheology has hitherto neglected to mention deformations other than viscous (e.g. plastic or brittle). However, we know that these deformations occur in the lithosphere, where the very idea of viscosity can lose its meaning in the vicinity of the surface. It is probable that the pitfall mentioned as a consequence of a strongly temperature-dependent viscosity reflects the limitation of this approach, which does not allow cold, viscous plates to be broken. Significant efforts have been made to mimic this effect while retaining a viscous formalism, notably by introducing the idea of a yield stress[26] (pseudo-plasticity), which has brought simulation results closer to the behavior of large-scale plates (e.g. Tackley 2000; Coltice et al. 2019) (Figure 1.10).

However, it seems that a potentially important aspect is missing from these models, the most successful in terms of surface expression of mantle dynamics, where plates emerge naturally from the flow. This is the tectonic inheritance of deformation, which sees certain sutures between plates on Earth reactivated during a subsequent tectonic cycle. It is easy to see why the notion of damage needs to be introduced. The nature of the property that carries this damage inheritance remains to be clarified (porosity, grain size). However, a simple formalism for this process has been successfully introduced by Bercovici and Ricard (2014), who show that such a concept could explain why Earth's plate tectonics do not seem to be observed as such on its cousin Venus, whose hotter surface would reduce its "memory" capacity[27].

---

26 The value favored for this yield stress by numerical models nevertheless remains lower than that observed in the laboratory for rocks.

27 This question is central, since explaining the dynamics of the Earth's mantle without explaining the different dynamics of the Venusian mantle seems rather futile. Another explanation for this dichotomy is the probable absence of water near the surface of Venus, the effect of which, although not mentioned in this all-too-brief write-up, is also identified as a first-order parameter for rock viscosity!

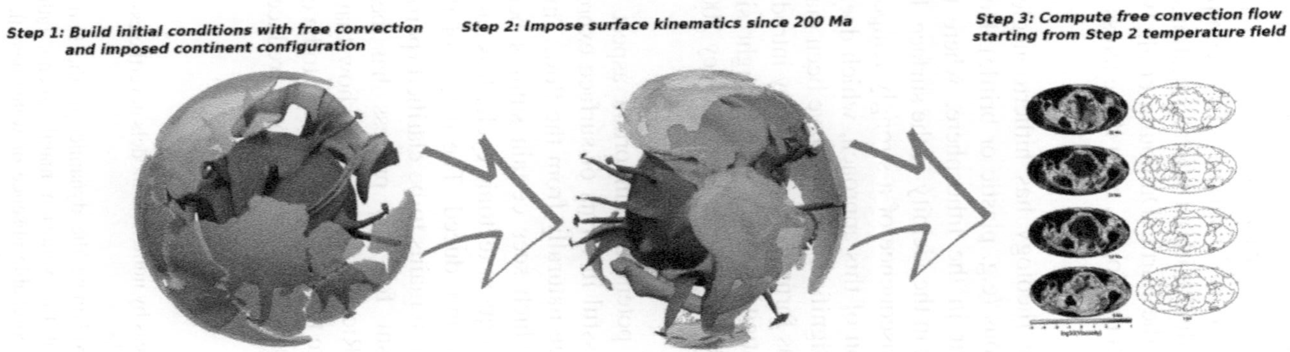

Figure 1.10. *Two examples of plate tectonic structures that can be reproduced with yield stress models. For a color version of this figure, see www.iste.co.uk/monteux/earth1.zip*

COMMENT ON FIGURE 1.10.– *(a) Initiation of rifting within a continent: colors indicate the depth of the base of the oceanic lithosphere (red to blue), the level of stress (gray level), and the amplitude of velocities whose orientation is represented by vectors. (b) Decomposition of the surface velocity field into divergence (blue-red) and vorticity (green-violet) whose amplitude demonstrates the obtaining of a toroidal field (see text). Figure taken from Coltice et al. (2019).*

1.4.4. *Composition*

Another essential aspect of convection in the Earth's mantle is the heterogeneity of rock composition. These are evident in the lithosphere (although the most obvious are in the crust, with a large-scale chemical dichotomy between continents and oceans, as well as variations in all smaller spatial scales, right down to the crystal scale). However, despite some uncertainty, a range of indirect measurements (primarily geochemical and seismological) bear witness to this for the mantle as a whole. These variations may concern trace elements (in very small quantities) or, conversely, major elements (mainly oxides). Trace elements have no influence on the physical characteristics of rocks, apart from radioactive elements, whose concentration determines the local heating rate. They are transported passively by mantle flow. The latter, on the other hand, affect properties such as density and viscosity, which are essential for convection. In practice, numerical models deal with the transport of these heterogeneities using tools such as PIC methods (see section 1.3.4), and may or may not include feedbacks with the density or viscosity field.

Convection models that take these heterogeneities into account, as they have been highlighted by other Earth science disciplines, are so numerous and varied that they cannot be listed in the format of this chapter. Nevertheless, we can mention a fruitful formalism proposed for convective regimes at the mantle scale, as evidenced by the identification of LLSVPs[28] by global seismic tomography from the 2000s onwards. If two immiscible fluids are juxtaposed in the same fluid layer, a composition field C (between 0 and 1) can translate their spatial distribution, whose 7evolution is described by a transport equation

28 Large low shear velocity provinces.

(formally, C replaces θ in equation [1.11] where the Laplacian[29] is omitted. The effect of composition on buoyancy leads to the introduction of a new dimensionless number:

$$\mathrm{Bu} = \frac{\Delta \rho_C}{\alpha \rho_0 \Delta T} \qquad [1.25]$$

where $\Delta \rho_C$ is the intrinsic density difference between the two fluids. The buoyancy term in equation [1.10] then becomes $\mathrm{Ra}\,(\theta + \mathrm{Bu}C)$ – very enlightening studies have been devoted to this type of convective dynamics.

The addition of a variable in composition is unfortunately not limited to the addition of a dimension in the space of fluid mechanics regimes obtained for mantle convection. The influence of composition on viscosity and density (which must necessarily be considered if the physics of mantle minerals is to be believed) means that a very wide range of models (and associated acronyms) can be produced, making a synthetic vision complex. This is the case of the example illustrated in Figure 1.11, which attempts to respond to seismological observations, still unclear, concerning the lower mantle.

1.4.5. *Several phases*

The heterogeneous composition of the mantle discussed above is essentially due to the melting/solidification of rocks. Crystallization of an ocean of magma occurs gradually, with different solidification temperatures for different crystals (see Chapter 3 for more details). Trace elements, for their part, have affinities for one or other of the liquid or solid phases, measured by a partition coefficient. This pattern, inherited from a hot birth for the Earth and telluric planets given the energetic events leading to their formation, naturally gives rise to variations in the composition of the rocks thus produced. The magmatism that follows in the course of geological evolution continues to produce variations in composition in a sort of opposite direction.

29 This equation does not include a diffusive term, given the weakness of chemical diffusivity when compared to thermal diffusivity.

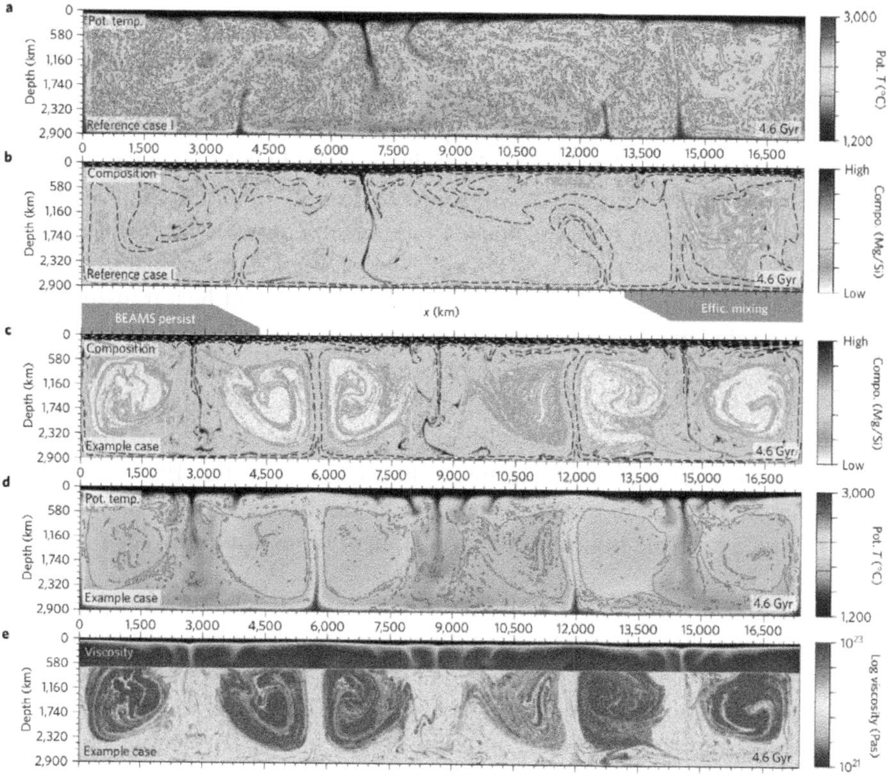

Figure 1.11. *Principle of BEAMS (Bridgmanite-enriched ancient mantle structures), silica-enriched lower mantle domains (low Mg/Si ratio) which, if associated with a higher viscosity (30 times that of the surrounding mantle), inhibit mixing over long geological timescales (c, d, e) when efficient mixing is produced without this increase in viscosity (a, b). For a color version of this figure, see www.iste.co.uk/monteux/earth1.zip*

COMMENT ON FIGURE 1.11.– *The temperature (a, d), composition (b, c) and viscosity (e) fields are shown. Figure taken from Ballmer et al. (2017).*

These phenomena also have an essential impact on convection in the Earth's solid mantle, both in terms of dynamic patterns and associated heat transfer. Studied in the greatest detail at the regional scale of oceanic ridge geodynamics (see, for example, the monograph by Katz (2022), which reveals its richness), complex effects of coupled rock-magma two-phase flow dynamics certainly come into play wherever solid and liquid coexist. In practice, if the rock is permeable, conservation equations are added to those for solid convection to describe the differential dynamics of the magma due to its inherent buoyancy and the compaction of the rock matrix, as well as melting/solidification phenomena. Porosity is introduced that quantifies the fraction of volume occupied by the molten liquid. In the case of oceanic mantle beneath ridges, both mechanical and chemical instabilities need to be taken into account to explain the formation of high-porosity bands, where magma flows rapidly upwards, resulting in highly efficient heat transfer. On a planetary scale, beyond the latent heat consumed or supplied during phase change, the extraction of large quantities of magma from the deeper mantle shapes a different thermal machine: Jupiter's moon Io, significantly larger than the Moon, seems to extract almost all its internal heat (more than twice that of the Earth) through volcanism. In the case of the Earth's mantle, volcanism or the emplacement of magma at the base of the crust may also have played a key role in heat transfer in ancient times (Lourenço et al. 2018).

In the case of magma oceans, the liquid layer is global on a planetary scale, inducing that the interface with the solid mantle (a boundary condition for the model) is also a phase change. This is particularly the case above a basal magma ocean (BMO) (Labrosse et al. 2007), whose thermal isolation guarantees its longevity: this condition at the lower, permeable boundary means that the convective patterns in the solid mantle are largely modified (Agrusta et al. 2020).

1.5. Conclusion

This chapter is only an overview. To conclude, mention may be made of the most recent numerical approaches that have been implemented in an attempt to produce predictive models for future mantle dynamics and plate tectonics through assimilation of past data, such as plate motions, and benefiting from mathematical tools introduced, for instance, in oceanography. For an example of this, see Bocher et al. (2018). Another aim is to make it possible to study a large number of models, taking advantage of developments in *machine learning* (see (Agarwal et al. 2020), for the evolution of the Mars mantle using neural networks).

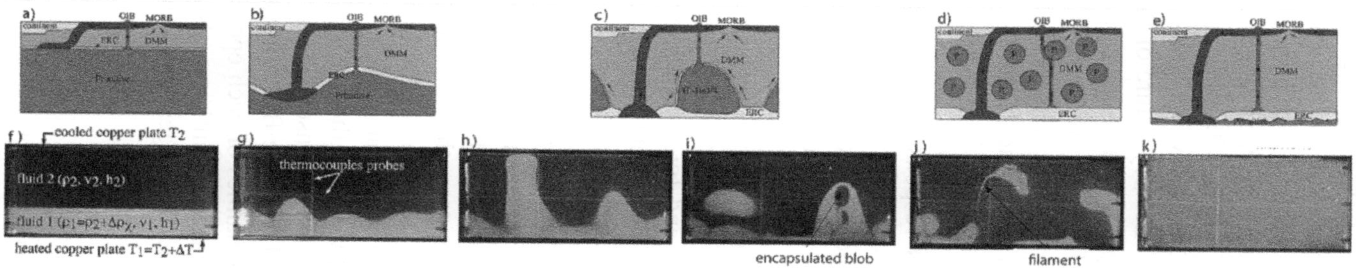

Figure 1.12. *Laboratory results for the dynamics of the Earth's mantle and major compositional reservoirs. The diagrams in the top row show the main families of dynamic models proposed at the time to reflect the lessons of geochemistry. The bottom row shows snapshots of the convection of two fluids of different density and viscosity: the evolution of the interface between the two fluids (the formation of a topography with variable topology, up to the most complete mixing) bears witness to each of the proposed models. Figure taken from Le Bars and Davaille (2004). For a color version of this figure, see www.iste.co.uk/monteux/earth1.zip*

One notable omission from the preceding pages concerns convection experiments conducted in the laboratory, where nature takes charge of solving the equations[30]: this approach has been particularly fruitful for the study of mantle dynamics, producing at each stage a counterpart to the numerical results (for convection of a fluid with variable viscosity and that of two layers of different composition, see (Davaille and Jaupart 1993; Le Bars and Davaille 2004)) (Figure 1.12). The very nature of these fluid mechanics experiments necessitates a conception that retains only the most salient aspects of the object whose model is being produced (the Earth's mantle), enabling a complete and enlightening physical analysis. Due to the inertia of the development of computational tools and hardware, these studies have also, more often than not, preceded numerical simulations.

Rooted in geology, since it is responsible for most of the observed phenomena at the Earth's surface (other contributions include the dynamics of its atmosphere and oceans, the evolution of climate, and life as an epiphenomenon), the dynamics of the Earth's mantle is a major physical problem. Its study has benefited from concomitant advances in many disciplines: materials science, mathematical applications, etc. The steps toward understanding it are fundamentally linked to the observations of geochemistry and geophysics (first and foremost seismology). This is why the Earth and planetary sciences are dependent on these data, since they have no other principle than to report on them.

In this context, the evolution of models of the Earth's mantle described in this chapter has followed a path that has seen studies move from general fluid mechanics principles, which account for a thermal machine that is certainly out of equilibrium but in successive states that only reflect an instantaneous balance, toward the inheritance that seems to need to be integrated to explain the current state. While the specific case of plate rheology has been mentioned, a more global, emblematic example would be the role of initial mantle melting, induced by primordial events and its major consequences on the heterogeneous composition (the LLSVPs inherited from the BMO!), which we do not seem to be able to omit to explain the dynamics of each epoch. From this point of view, models of the Earth's and planets' mantle are also akin to a historical discipline that must bear witness to the stages, and sometimes events, that shape its evolution.

30 With the occasional ingenious use of a microwave oven to mimic the internal heating of the fluid (Limare et al. 2015).

1.6. References

Agarwal, S., Tosi, N., Breuer, D., Padovan, S., Kessel, P., Montavon, G. (2020). A machine-learning-based surrogate model of Mars' thermal evolution. *Geophys. J. Int.*, 222(3), 1656–1670.

Agrusta, R., Morison, A., Labrosse, S., Deguen, R., Alboussière, T., Tackley, P.J., Dubuffet, F. (2020). Mantle convection interacting with magma oceans. *Geophys. J. Int.*, 220(3), 1878–1892.

Ballmer, M.D., Houser, C., Hernlund, J.W., Wentzcovitch, R.M., Hirose, K. (2017). Persistence of strong silica-enriched domains in the Earth's lower mantle. *Nat. Geo.*, 10(3), 236–240.

Bénard, H. (1901). Les tourbillons cellulaires dans une nappe liquide. Méthodes optiques d'observation et d'enregistrement. *J. Phys. Theor. Appl.*, 10(1), 254–266.

Bénard, H. and Avsec, D. (1938). Travaux récents sur les tourbillons cellulaires et les tourbillons en bandes. Applications à l'astrophysique et à la météorologie. *J. Phys. Radium*, 9(11), 486–500.

Bercovici, D. and Ricard, Y. (2014). Plate tectonics, damage and inheritance. *Nature*, 508(7497), 513–516.

Bocher, M., Fournier, A., Coltice, N. (2018). Ensemble Kalman filter for the reconstruction of the Earth's mantle circulation. *Nonlin. Proc. Geophys.*, 25(1), 99–123.

Choblet, G., Čadek, O., Couturier, F., Dumoulin, C. (2007). OEDIPUS: A new tool to study the dynamics of planetary interiors. *Geophys. J. Int.*, 170, 9–30.

Coltice, N., Husson, L., Faccenna, C., Arnould, M. (2019). What drives tectonic plates? *Sci. Adv.*, 5(10), eaax4295.

Davaille, A. and Jaupart, C. (1993). Transient high-Rayleigh-number thermal convection with large viscosity variations. *J. Fluid Mech.*, 253, 141–166.

Dumoulin, C., Bercovici, D., Wessel, P. (1998). A continuous plate-tectonic model using geophysical data to estimate plate-margin widths, with a seismicity-based example. *Geophys. J. Int.*, 133(2), 379–389.

Gerya, T.V. and Yuen, D.A. (2003). Characteristics-based marker method with conservative finite-differences schemes for modeling geological flows with strongly variable transport properties. *Phys. Earth Plan. Int.*, 140, 295–320.

Haskell, N.A. (1935). The motion of a viscous fluid under a surface load. *Physics*, 6, 265–269.

Heister, T., Dannberg, J., Gassmöller, R., Bangerth, W. (2017). High accuracy mantle convection simulation through modern numerical methods? II: Realistic models and problems. *Geophys. J. Int.*, 210(2), 833–851.

Holmes, A. (1931). XVIII. Radioactivity and earth movements. *Transactions of the Geological Society of Glasgow*, 18(3), 559–606.

Ismail-Zadeh, A. and Tackley, P. (2010). *Computational Methods for Geodynamics*. Cambridge University Press, Cambridge.

Katz, R.F. (2022). *The Dynamics of Partially Molten Rock*. Princeton University Press, Oxford.

Kervazo, M., Tobie, G., Choblet, G., Dumoulin, C., Běhounková, M. (2021). Solid tides in Ios partially molten interior: Contribution of bulk dissipation. *Astron. Astrophys.*, 650, A72.

Kramer, S.C, Davies, D.R., Wilson C.R. (2021). Analytical solutions for mantle flow in cylindrical and spherical shells. *Geosci. Model Dev.*, 14, 1899–1919.

Krishnamurti, R. and Howard, L.N. (1981). Large-scale flow generation in turbulent convection. *PNAS*, 78(4), 1981–1985.

Labrosse, S., Hernlund, J.W., Coltice, N. (2007). A crystallizing dense magma ocean at the base of the Earth's mantle. *Nature*, 450(7171), 866–869.

Le Bars, M. and Davaille, A. (2004). Whole layer convection in a heterogeneous planetary mantle. *J. Geophys. Res.*, 109(B3), B03403.

Lehmann, I. (1936). P. Bureau Central Sismologique International, Strasbourg, 14, 87–115.

Limare, A., Vilella, K., Di Giuseppe, E., Farnetani, C.G., Kaminski, E., Surducan, E., Neamtu, C., Fourel, L., Jaupart, C. (2015). Microwave-heating laboratory experiments for planetary mantle convection. *J. Fluid Mech.*, 777, 50–67.

Lourenço, D.L., Rozel, A.B., Gerya, T., Tackley, P.J. (2018). Efficient cooling of rocky planets by intrusive magmatism. *Nat. Geo.*, 11(5), 322–327.

Machetel, P., Rabinowicz, M., Bernadet, P. (1986). Three-dimensional convection in spherical shells. *Geophys. Astrophys. Fluid Dyn.*, 37, 57–84.

McKenzie, D.P., Roberts, J.M., Weiss, N.O. (1974). Convection in the Earth's mantle: Towards a numerical solution. *J. Fluid Mech.*, 62, 465–538.

Oreskes, N. (1999). *The Rejection of Continental Drift: Theory and Method in American Earth Science*. Oxford University Press, Oxford.

Parmentier, E.M. and Sotin, C. (2000). Three-dimensional numerical experiments on thermal convection in a very viscous fluid: Implications for the dynamics of a thermal boundary layer at high Rayleigh number. *Phys. Fluids*, 12, 609–617.

Ribe, N.M. (2015). Analytical approaches to mantle dynamics. In *Treatise on Geophysics*, volume 7, Bercovici, D. (ed.). Elsevier, Amsterdam.

Ricard, Y. (2015). Physics of mantle convection. In *Treatise on Geophysics*, volume 7, Bercovici, D. (ed.). Elsevier, Amsterdam.

Schubert, G., Turcotte, D.L., Olson, P. (2001). *Mantle Convection in the Earth and Planets*. Cambridge University Press, Cambridge.

Solomatov, V.S. (1995). Scaling of temperature- and stress-dependent viscosity convection. *Phys. Fluids*, 7(2), 266–274.

Strutt, J.W. (1916). On convective currents in a horizontal layer of fluid when the higher temperature is on the under side. *Philosophical Magazine*, 32, 529–546.

Tackley, P.J. (2000). Self-consistent generation of tectonic plates in time dependent, three-dimensional mantle convection simulations. 1. Pseudoplastic yielding. *Geochem. Geophys. Geosys*, 1(8), 2000GC000036.

Tackley, P.J. (2008). Modeling compressible mantle convection with large viscosity contrasts in a three-dimensional spherical shell using the Yin-Yang grid. *Physics of the Earth and Planetary Interiors*, 171, 7–18.

Torrance, K.E. and Turcotte, D.L. (1971). Thermal convection with large viscosity variations. *J. Fluid Mech*, 47, 13–125.

Turcotte, D.L. and Oxburgh, E.R. (1967). Finite amplitude convective cells and continental drift. *J. Fluid Mech*, 28, 29–42.

Wegener, A. (1915). *Die Enstehung der Kontinente und Ozeane*. Vieweg, Braunschweig.

White, D.B. (1988). The planforms and onset of convection with a temperature-dependent viscosity. *J. Fluid Mech.*, 191, 247–286.

Zhong, S.J., Zuber, M.T., Moresi, L.N., Gurnis, M. (2000). Role of temperature dependent viscosity and surface plates in spherical shell models of mantle convection. *J. Geophys. Res.*, 105, 11063–11082.

2

How the Earth's Core and Mantle are Separated: Geochemical and Dynamic Constraints

Julien MONTEUX and Maud BOYET
LMV, CNRS, Université Clermont Auvergne, IRD, OPGC, Clermont-Ferrand, France

2.1. Introduction

The separation of the Earth's core and mantle is a highly energetic process that took place during or just after the Earth's major accretionary stages (Stevenson 1981). Thus, this primitive phase of the Earth is probably intimately linked to impact processes. On collision with an impactor with a radius greater than 100 km, the energy supplied by the collision can be added to the fossil energy derived from the radioactivity of short-lived elements. This can generate melting zones on varying scales, within which the iron from the impactor is dispersed in the form of droplets. During the primitive stages of planet formation, impacts may have caused local segregation between dense iron and lighter silicates in the heated zone (Tonks and Melosh 1992; Deguen et al. 2011). Once gathered at the base of the impact-melted reservoir, large volumes of iron could then have migrated toward the center of the Earth and contributed to the formation of the Earth's core by percolation (Shannon and Agee 1998) or by diapirism (Stevenson 1981).

Structure and Dynamics of the Earth's Interior 1,
coordinated by Julien MONTEUX.

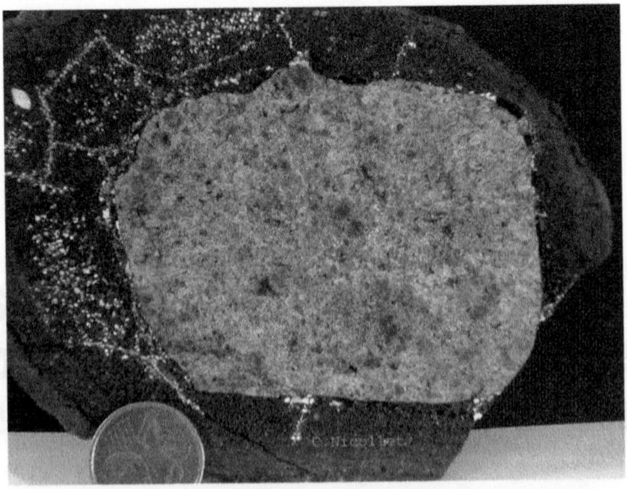

Figure 2.1. *Mantelic xenolith trapped inside a volcanic bomb (credits: http://christian.nicollet.free.fr/). For a color version of this figure, see www.iste.co.uk/monteux/earth1.zip*

Access to samples of the Earth's mantle is difficult from the surface. However, lavas with high ascent velocities can tear off pieces of mantle as basalts rise. These mantle fragments, known as xenoliths, can become trapped inside volcanic bombs (Figure 2.1). Such samples can also be found locally on the continental crust in tectonically exhumed portions of oceanic lithosphere: ophiolites. Geochemical analysis of these mantle samples reveals two important points. First, the concentration of siderophile elements in the upper mantle is ~ 100 times lower than in the meteorites that may have served as building materials for the planets. On the other hand, the concentration of highly siderophile elements (HSE) in these same samples is ~ 100 times higher than predicted by models calculated from metal/silicate partition coefficients determined by experiments (Becker et al. 2006; Lorand et al. 2008, 2012).

The degree of chemical equilibrium between the mantle and the Earth's core depends on the type of differentiation and, in particular, the characteristic volumes of iron (centimetric droplets or multi-kilometer diapirs) brought into play after impact between two differentiated bodies. Indeed, if the characteristic volumes of iron involved in core formation are large (kilometric diapirs), separation rates will be very rapid and thermo-chemical equilibrium between the metallic and silicate phases will be partial. On the other hand, if the iron/silicate separation takes place via centimetric iron droplets, exchanges between the two phases will be facilitated and thermo-chemical equilibrium will be achieved. Out-of-equilibrium differentiation is one hypothesis put forward to explain the anomalous concentrations of HSE in terrestrial rocks (Kleine and Rudge 2011). Thus, the chemical composition of the mantle of telluric planets provides us with information on the characteristics of core/mantle differentiation.

The iron/silicate separation processes in the planet's interior and the characteristic size of the iron flow in the mantle are governed by the viscosities involved, whether in the metallic or silicate phase. In a turbulent magma ocean, iron is easily fragmented into centimetric droplets before accumulating at its base. Beneath the magma ocean, the mantle may be in a solid state, slowing down the separation of iron from silicates (Figure 2.2). The iron thus accumulates in a layer at the interface between the low-viscosity magma ocean and the highly viscous solid mantle. Subsequently, iron diapirs whose size depends on the viscosity of the underlying material are mobilized by gravitational instabilities and plunge toward the center of the planet (Figure 2.2). The viscosity profile in the Earth's interior is intimately linked to the distribution of heat at depth. This distribution can be strongly influenced by the way iron migrates toward its center (Ke and Solomatov 2009). When a dense volume of iron falls, potential energy is dissipated as heat in the mantle or core. If the iron plunges through long, narrow channels, the dissipated heat is mainly found in the core. However, if the iron plunges in the form of large drops or diapirs, the potential energy is dissipated mainly in the mantle. Depending on the segregation mechanisms, this gives rise to different thermal profiles, with either a core much hotter than the mantle, or a very hot, sometimes even molten, mantle. The methods of differentiation, and in particular the characteristic volumes of iron (centimetric droplets or plurikilometric diapirs) involved in the iron/silicate separation, therefore condition the primitive thermal state of the Earth's mantle. Samuel et al. (2010) have also shown that a large number of small iron diapirs plunging into a magma ocean favors iron/silicate heat exchange and leads to a relatively hot core, as on Earth. A contrario, a small number of large diapirs has the opposite effect and could lead to a core with a relatively moderate temperature. In this scenario, which could correspond to what happened on Mars, a dynamo would be present early in the planet's history, but only for a short time. Iron/silicate separation by diapirism tends to leave the deep mantle significantly hotter than its upper part, which could also generate a melting zone above the core/mantle boundary.

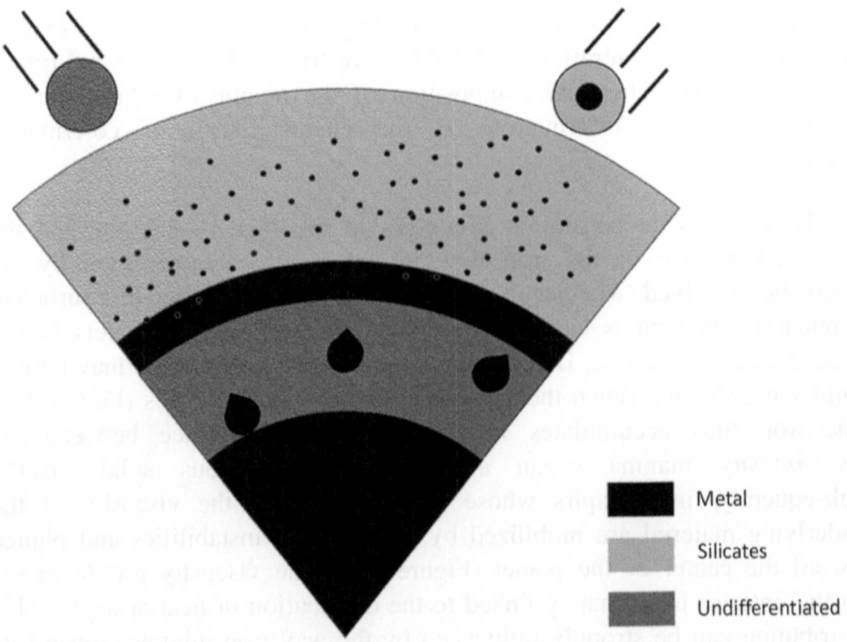

Figure 2.2. *Model of the formation of the Earth's core from a deep magma ocean (according to Wood (2011)). For a color version of this figure, see www.iste.co.uk/monteux/earth1.zip*

In this chapter, we discuss the dynamics of core/mantle separation and the chemical and thermal consequences of this major episode in the Earth's evolution. First, we provide an overview of the Earth's chemical structure, showing in particular how the modalities of core/mantle separation may have affected the chemical composition of its mantle. We will then present the different iron/silicate separation processes and the role played by giant impacts in these dynamics. In this chapter, we will focus primarily on the case of the Earth, but we will also open the discussion to other telluric planets.

2.2. How the Earth's core and mantle are separated: geochemical clues

The metallic phase is mainly found in the Earth's core, while silicates are the main chemical component of the overlying mantle. While the internal structure of the Earth and its telluric planets is now increasingly well understood, the modalities of core/mantle separation still raise many questions.

2.2.1. *Internal composition model of the Earth and telluric planets*

The Earth is divided into distinct physical units, with a solid metallic inner core surrounded by a liquid metallic outer core. The core is enveloped by a silicate mantle, followed by a crust (oceanic or continental). The Earth's surface is characterized by oceans and a gaseous atmosphere (Figure 2.3).

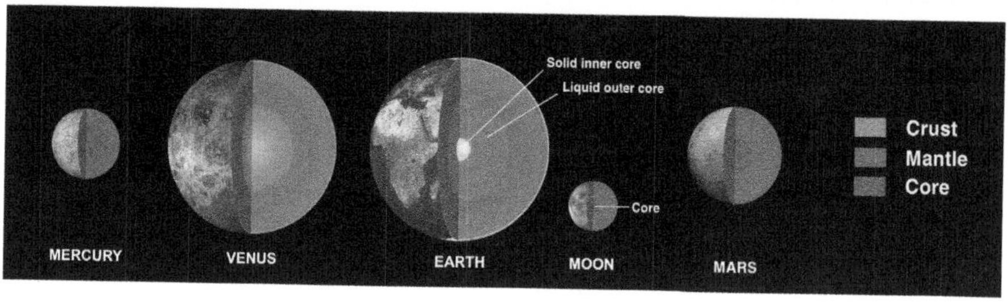

Figure 2.3. *Schematic representation of the current internal structure of the terrestrial planets and the Moon (credit: NASA). For a color version of this figure, see www.iste.co.uk/monteux/earth1.zip*

COMMENT ON FIGURE 2.3.– *In this figure, the planets are classified according to their heliocentric distance, and the size scales are respected. The figure illustrates the disparity in core/mantle ratios among the telluric planets. Indeed, while the core would represent ~1% of the Moon's volume, it could represent ~60% of Mercury's volume. For the Earth, the ratio is 17%. This disparity illustrates the variety of planetary evolutions during the early stages of their formation.*

Petrological and geochemical data now enable us to better constrain the physical properties and chemical composition of the Earth's major reservoirs. The Earth's internal structure is mainly known from seismic data, which enable us to observe the boundaries between terrestrial reservoirs indirectly. As seismic wave velocities are a function of the density of the materials crossed, it is possible to use inversion models to get an idea of the chemical composition of deep geological reservoirs (Mattern et al. 2005). Seismic tomography is thus an essential tool for obtaining an image of the Earth's internal structure (PREM model). Space missions have made it possible to set up seismic measurement tools on the Moon (Apollo 17 mission) and Mars (InSight mission). Although the resolution of the data is lower than on Earth due to the small number of seismometers deployed, the data obtained are nonetheless essential for better constraining the internal structure of these two objects (e.g. Weber et al. 2011; Samuel et al. 2023). Mars would thus have a metallic core with a radius of $\sim 1,800$ km, whereas the radius of the lunar core is estimated at between 250 and 400 km.

The internal structures of other planets are estimated by studying their moments of inertia and magnetic fields. The Solar System's telluric objects (Mercury, Venus, Mars and Moon) all have a denser metallic core (Figure 2.3). Of these objects, we only have samples from the upper envelopes of the Earth, Moon (Apollo missions and rare lunar meteorites) and Mars (Shergottite–Nakhlite–Chassignite (SNC) meteorites).

2.2.2. *Chemical elements and their properties*

The Earth is a differentiated planet, with envelopes of different chemical compositions. The chemical elements of the periodic table can be classified according to their chemical affinities, which determine their distribution in the major terrestrial reservoirs. Four groups have been defined by V. Goldschmidt (Figure 2.4):

– Atmophile elements have a particular affinity for the gas phase and will concentrate preferentially in the atmosphere.

– Lithophile elements are concentrated in the silicate parts of planets (mantle, crust).

– Chalcophile elements have a strong affinity for sulfide phases.

– Siderophile elements have a strong affinity for the metallic phase. They are depleted in the silicate parts of differentiated bodies as they migrate into planetary cores during iron/silicate differentiation.

The chemical behavior of the elements and the compositions of the samples can thus be used to constrain the differentiation processes. For example, the concentration of siderophile elements measured in samples of the Earth's mantle can be traced back to the conditions under which the metallic and silicate phases separated. The parameters that modify elemental behavior are pressure, temperature, oxygen fugacity and chemical composition.

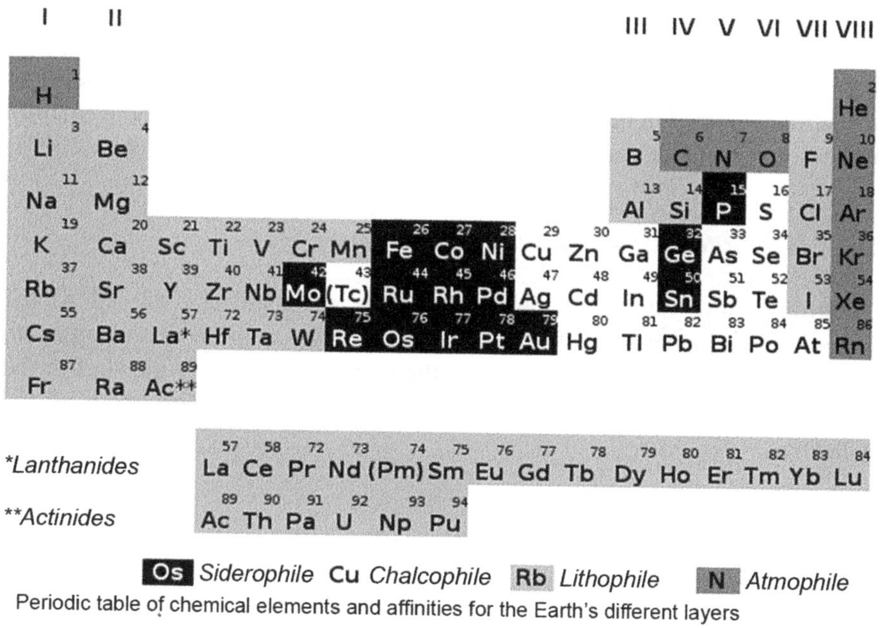

Figure 2.4. *Goldschmidt classification in the periodic table of chemical elements (credit: planet-terre.ens-lyon.fr)*

The propensity of certain elements to distribute themselves in different reservoirs can be used to date an episode of differentiation. Radioactive decay can be used to date an event that leads to the chemical fractionation of father and son elements. Several isotopic systems, such as the U-Pb system, are fractionated during iron/silicate differentiation. During the formation of the Earth's core, uranium and lead were fractionated, leaving a uranium-enriched mantle that produced more radiogenic lead. When measurements of terrestrial samples (lavas, sediments) are plotted in the $^{207}Pb/^{204}Pb = f(^{206}Pb/^{204}Pb)$ diagram, they align with a 4.45 Ga isochron. This means that the separation of the Earth's core occurred before 100 million years after the start of the formation of the Solar System.

The extinct radioactivity ^{182}Hf-^{182}W characterized by a half-life of 9 million years (period during which 50% of the quantity of the father isotope is decayed) also provides strong constraints on the chronology of the Earth's differentiation. Hafnium (Hf) is a lithophile element that does not enter the metallic phase, while tungsten (W) is a siderophile element. If iron/silicate separation occurs during the lifetime of ^{182}Hf, the silicate reservoir, characterized by a high Hf/W ratio, will acquire an excess of radiogenic ^{182}W over the chondrites, which represent undifferentiated material. Terrestrial rocks do indeed have an excess of ^{182}W over chondrites, proving that core formation occurred in the first 60 million years after the start of the formation of the Solar System (Yin et al. 2002; Kleine et al. 2002; Touboul et al. 2007).

2.2.3. *Meteorites: Building blocks of the Earth*

Before going into more detail about the mechanisms that led to the formation of planetary cores, we need to give a brief introduction to meteorites. These objects tell us about the materials present at the start of the formation of the Solar System. These contributed to the formation of the planets.

2.2.3.1. *Undifferentiated meteorites: chondrites*

The vast majority of meteorites that fall and are sampled on Earth are chondrites. Chondrites are rocky bodies made up of (Figure 2.5):

1) refractory inclusions rich in Ca and Al (CAI). They correspond to the first phases condensed during the cooling of the protosolar nebula;

2) chondrules, spherules composed of silicate minerals that account for up to 80% of a chondrite's volume – these elements are droplets of molten primordial matter;

3) metal nodules and sulfides;

4) a silicate matrix acting as cement.

The abundance of these components varies according to the type of chondrite. Among these chondrites, there are three main groups: ordinary chondrites, enstatite chondrites and carbonaceous chondrites.

To date these objects, it is possible to use absolute chronology methods based on long-period radioactivity. For example, the Pb-Pb method is based on determining the isotopic composition of lead, two of whose isotopes, ^{206}Pb and ^{207}Pb, derive in part from the natural decay of two long-lived radioactive isotopes of uranium: ^{238}U and ^{235}U. As these decays involve very long half-lives (greater than 4.468 Ga), it is possible to use these systems to date primitive events and objects. Absolute radiometric ages thus indicate that chondrites are composed of structures of very ancient ages. In 1956, C. Patterson carried out the first dating of a meteorite using the Pb-Pb method. He obtained an age of between 4.49 and 4.55 Ga, depending on the radioactive constants available. Among chondrite constituents, CAIs are the oldest, with absolute ages of $\sim$ 4,567 million years (Amelin et al. 2002; Bouvier et al. 2008; Connelly et al. 2012). Since these objects were the first to condense during the cooling of the proto-solar nebula, they correspond to the time t_0 considered as the age of formation of the Solar System.

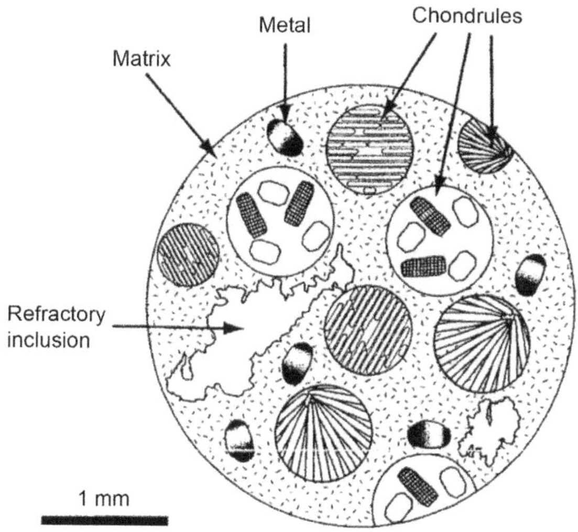

Figure 2.5. *Schematic of chondritic meteorite composition (from Beck (2005))*

Despite the advent of highly sophisticated geochemical analysis tools, the low decay rates of long-period radioactivities limit the precision of analyses and the development of a fine chronology between very ancient events. So, to obtain precise relative ages between events and to date the formation of the first solids in the Solar System, geochemists also use short-period radioisotopes, which offer high temporal resolution. Among these tools, the most widely used is based on the decay of ^{26}Al into ^{26}Mg with a half-life of 0.73 million years. These extinct radioactivities existed only during the first million years of the Solar System. By coupling the ages obtained by the U-Pb method and the extinct radioactivities, the chronology of the Solar System is established. The process by which chondrules are formed remains the subject of much debate. Chondrules are slightly younger than CAIs. The study of short- and long-lived radio chronometers indicates that chondrules form between $\sim$ 1 Myr and $\sim$ 4 Myr after CAIs (Amelin et al. 2002; Krot et al. 2005; Villeneuve et al. 2009). The formation of these rapidly solidified droplets of primitive matter could be linked to a decompression episode following a shock wave, electric discharges, and/or irradiation in the vicinity of the proto-Sun. The formation of planetesimals brought together chondrules and silicate-iron dust in hectometric to kilometer-sized bodies. Chondrites come from asteroids too small to have been melted. More recently, they have been torn from parent bodies (small asteroids) by inter-asteroidal collisions, somewhere in the asteroid belt located between 2.06 and 3.27 astronomical units (AU) between Mars and Jupiter.

2.2.3.2. *Differentiated meteorites*

Differentiated meteorites come from bodies that have undergone large-scale chemical differentiation. These meteorites thus originate from the fracturing of larger original bodies made up of various envelopes, as identified for the Earth's internal structure. Differentiated meteorites are much less abundant than undifferentiated meteorites. In fact, they account for only 30% of meteorites recorded to date. Differentiated meteorites differ in their petrogenetic origins and in their proportions of iron and silicates. Among these meteorites, three groups can be distinguished:

1) achondrites, which are silicate rocks derived from the crust of parent bodies (HED, SNC, angrites, ureilites and aubrites);

2) siderites, which are composed of over 90% Fe-Ni metal and probably originate from the core of parent bodies;

3) iron–silicate associations (pallasites and mesosiderites). They are possible relics of the interface between metallic core and silicate mantle. These meteorites are known as lithosiderites or siderolites.

Like the Earth, the parent bodies of differentiated meteorites underwent chemical differentiation, with the denser elements migrating toward the center of the protoplanet. The study of differentiated meteorites thus provides information on the iron/silicate separation processes within the parent bodies from which they originate. However, the differentiation processes take place under different conditions, since the bodies are characterized by varying sizes and chemical compositions, which will modify the conditions of pressure, temperature and oxygen fugacity. The study of iron meteorites provides an insight into metal/silicate differentiation, but the pressure conditions are likely to be different from those under which the Earth's core was formed.

2.2.3.3. *What are the Earth's building materials?*

Analysis of meteorite composition provides crucial information on the composition of the first planets to form around the Sun. These first objects formed from a proto-planetary disk which, early in the history of the Solar System, displayed chemical, isotopic and mineralogical zonations. Such zonations are found today in the different classes of primitive meteorites (chondrites and differentiated meteorites) recorded on Earth, but also in the distribution of asteroids in the main belt. This zonation would thus reflect the primitive zonation of the Solar System, with E- and S-type asteroids in the inner parts, possible parent bodies of enstatite chondrites and ordinary chondrites, respectively. C-type asteroids, the parents of carbonaceous chondrites, would be concentrated in the outermost parts of the asteroid belt (DeMeo and Carry 2014).

To determine as precisely as possible which type of meteorite may have served as building material for the Earth, geochemists compare the isotopic signatures of meteorites with those of terrestrial samples. Of the meteorites sampled, enstatite chondrites show the most similarities to Earth (Javoy et al. 2010). Enstatite chondrites and Earth show similar compositions for many isotopic systems such as O, Ca, N, Mo, Ru, Os, Cr, Ni, Ti and Sr (Javoy et al. 2010; Dauphas 2017). However, these chondrites have a lower Mg/Si ratio than Earth's upper mantle. The Earth also has higher abundances of refractory lithophile elements and lower concentrations of moderately volatile elements (such as the alkaline elements Li, Na, K and Rb) compared with enstatite chondrites. To explain these differences, Boujibar et al. (2015) have proposed a model involving the erosion of 15% of the early Earth's mass by multiple collisions to reconcile the compositional differences between enstatite chondrites and the Earth. In effect, the loss of the Earth's proto-crust through impacts raises the Earth's Mg/Si ratio to its present value. To match all major element compositions, the model proposed by Boujibar et al. (2015) implies preferential loss of volatile lithophile elements and recondensation of refractory lithophile elements after impacts.

Some chemical elements have the particularity of changing affinity according to pressure, temperature and oxygen fugacity conditions. This is the case for niobium (Nb) and tantalum (Ta), which are among the lithophile elements in Goldschmidt's classification (Figure 2.4). However, Cartier et al. (2014) have shown that these two elements both become siderophile elements for reducing conditions, that is, at low oxygen fugacity. By applying their partition coefficient data in theoretical accretion models, they confirmed that planetary accretion of reduced materials (such as enstatite chondrites) played a major role in the chemical evolution of the early Earth. The final stages of accretion involved more oxidized material (ordinary chondrites). Heterogeneous accretion with a contribution of more oxidized material in the later phases helps explain many of the Earth's geochemical features (Rubie et al. 2011).

2.2.4. *Constraints using moderately siderophile elements*

While the sampling and chemical analysis of the lower mantle are today difficult to envisage, the chemical composition of the Earth's upper mantle, for which reliable and precise data have been obtained, provides important information on the accretionary history of the Earth and the planets of the inner Solar System. During the formation of the core of telluric planets, iron and silicates separate to form the two major reservoirs that are the core and the mantle. During this separation, siderophile elements are naturally attracted to iron and tend to concentrate in the core, while lithophile elements are trapped in the silicate mantle (Figure 2.4). The analysis of siderophile element concentrations in upper mantle samples can therefore provide vital information on how the Earth's core was formed.

Among siderophile elements, moderately siderophile elements (Figure 2.6) are essential geochemical tracers for characterizing core/mantle separation modalities. Partition coefficients $D^{met/sil}$ characterize the propensity of an element to concentrate in the metallic or silicate phase. Generally speaking, the partition coefficients of moderately siderophile elements vary greatly with pressure, temperature and/or oxygen fugacity, to a greater or lesser degree depending on the element. The $D^{met/sil}$ values are therefore robust constraints for defining core/mantle separation modalities.

Among the moderately siderophile elements, nickel and cobalt provide major constraints on the pressure conditions that prevailed when these two reservoirs were separated, as their $D^{met/sil}$ partition coefficient varies by three orders of magnitude between 1 bar and 100 GPa (Chabot et al. 2005). The composition of these two elements is depleted in the mantle compared with a chondritic composition, but their concentration remains higher than predicted by low-pressure experiments, leading to the paradox of an excess of Ni and Co in the mantle. High-pressure and high-temperature experiments have explained this excess, leading to the proposal that core/mantle separation occurred under pressure and temperature conditions of between 25 and 60 GPa and 2,200 and 4,000 K, respectively (Righter and Drake 1999; Gessmann and Rubie 2000; Li and Agee 2001; Bouhifd and Jephcoat 2011). These conditions are compatible with an iron/silicate separation within a magma ocean with a depth of between 800 and 1,500 km.

2.2.5. *Excessive concentration of HSE*

We saw earlier that the mantle was depleted of moderately siderophile elements by the formation of the core. HSE are more depleted than moderately siderophile elements in the mantle because their affinity for the metal phase is stronger, that is, they have even higher metal/silicate partition coefficients. Despite this, HSE remain in excess in the upper mantle compared to equilibrium and low pressure differentiation of chondritic material (Bézos et al. 2005; Becker et al. 2006; Lorand et al. 2008, 2012) (Figure 2.6).

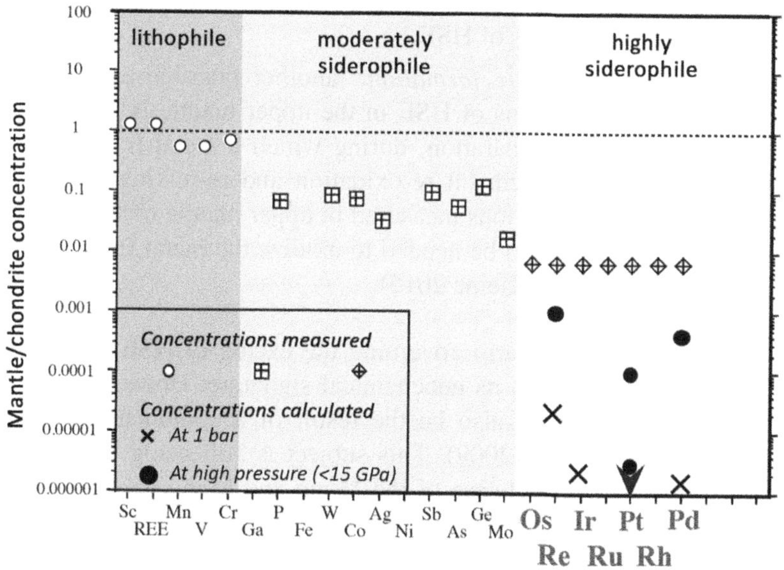

Figure 2.6. *Mantle elemental abundances relative to a carbonaceous chondrite (from Lorand et al. (2008)). For a color version of this figure, see www.iste.co.uk/monteux/earth1.zip*

To explain this peculiarity, the following three hypotheses can be invoked (Wood et al. 2006):

1) *A late accretion*: a sprinkling of material during an accretion episode occurring after the formation of the Earth's core would have brought siderophile elements into the mantle. This late veneer accretion phase would have added around 0.5% of chondritic material. The added siderophile elements would thus have been unable to reach the core during differentiation and would now be trapped in the mantle.

2) *A high-pressure core formation*: the iron/silicate partition coefficients of certain siderophile elements are sensitive to pressure and temperature conditions. Thus, high pressure and temperature conditions could explain why HSE would be in excess in the upper mantle. But to obtain such conditions, it is necessary to invoke a very deep magma ocean (between 700 and 1200 km) at the base of which the iron accumulated and then equilibrated with the surrounding silicates. The iron then joined the metallic core in the form of diapirs in the growing core. This high-pressure scenario can explain the distribution of many elements between core and mantle without invoking assumptions of particular conditions on the oxidation state of the growing Earth. However, a late veneer like the one mentioned above is still needed to explain the concentrations of HSE.

3) *A non-equilibrium core formation*: another possible explanation for the anomalous concentrations of HSE in the upper mantle is a more or less catastrophic iron/silicate separation, during which a metal fraction remains mixed in the mantle. Subsequent re-oxidation and re-mixing of this metal would lead to the concentrations measured in upper mantle rocks. However, a huge amount of water would be needed to oxidize the metal fraction trapped in the mantle (Nimmo and Kleine 2015).

It is likely that the scenario governing the excess of HSE in the Earth's upper mantle is recorded in its geochemical signature. However, the excess concentration of HSE may also be the result of a combination of several of these scenarios (Walker 2009). This subject is still being debated in the scientific community. In the case of the Moon and Mars, we have no direct access to the HSE composition of the mantle, but they can be recalculated from fusion products. Geochemical analysis of SNC meteorites and lunar samples suggests a greater mass of material accreted during the late veneer for Earth compared with Mars (10 times less) and the Moon (1,000 times less) (Bottke et al. 2010). The difference in HSE composition between Earth, Mars and the Moon probably illustrates differences in the modalities of formation and primitive evolution between these telluric bodies.

2.3. Iron/silicate separation in a magma ocean

The degree of chemical equilibrium between the mantle and the Earth's core depends on the type of differentiation and, in particular, the characteristic volumes of iron (centimetric droplets or plurikilometric diapirs) (Rubie et al. 2003; Dahl and Stevenson 2010). Thus, measurements of the chemical and isotopic composition of the mantle of telluric planets provide constraints on the migration process of the metallic phase through the mantle to form the core.

Iron/silicate separation processes are closely linked to the evolution of the early Earth's thermal profile. During its accretion, the young Earth became progressively warmer, notably through the combined action of impacts, which dissipate their kinetic energy in the form of heat, and radioactive heating linked to the decay of short-period elements such as ^{26}Al or ^{60}Fe (see Chapter 3). These processes caused progressive heating, particularly in the outer parts of the Earth. Kaula (1979) showed in particular that the accumulation of primitive heat by accretion could form magma oceans. According to these models, the magma ocean stage seems inevitable for an Earth formed in less than 50 million years, provided that at least 10% of the kinetic energy of the impactors is converted into heat. Thus, core/mantle differentiation may have involved both iron/silicate separation processes where both phases were liquid, and segregation processes where the silicate phase was solid. In the former case, separation occurs over short time scales and rather small characteristic lengths. In the second case, separation times are longer and the volumes involved are greater (Figure 2.2).

Core-mantle separation took place early in the Earth's history (Kleine et al. 2002), a period during which meteorite impacts involving planetesimals or proto-planets were frequent (Agnor et al. 1999). Core formation is therefore probably strongly linked to impact processes during planet accretion. These impacts may have brought large volumes of iron or the energy needed to separate iron from silicates into the impacted body. During terrestrial accretion, many impactors collided while the mantle was significantly melted. In such a context, the metallic phase of the impactor would disperse into the less dense magma ocean and then plunge in the form of an iron rain scenario (Figure 2.2). During this process, two fragmentations take place: a fragmentation directly during the impact (Kendall and Melosh 2016), then a fragmentation in the molten reservoir (Wacheul et al. 2014).

2.3.1. *Fragmentation on impact*

During an impact involving two differentiated objects, the impactor will undergo intense deformation, leading not only to the fragmentation of its silicate mantle, but also of its metallic core. Genda et al. (2017) have modeled the consequences of an impact between the Earth and a differentiated impactor ($R = 1,500$ km) after the formation of the Earth–Moon system. For a statistically probable oblique collision (i.e. with an impact angle of 45°), the core of the impactor elongates and then disintegrates into fragments of decametric size. Some of these fragments hit the Moon in orbit and contribute to the formation of the lunar core, but most fall back to Earth as secondary impactors. Genda et al. (2017) show that a single oblique impactor can supply sufficient quantities of HSE to primordial terrestrial silicate reservoirs in the early Hadean.

The impact process also stretches the metallic phase, promoting the dispersion of liquid iron into a much larger volume in the impact-melted zone. Kendall and Melosh (2016) have developed numerical impact models (see Chapter 3) incorporating numerical tracers that track the stretching of the impactor's metallic phase during the collision. The evolution of the distance between the tracers provides a measure of this core stretching. Stretch statistics imply the dispersion of the iron core with an initial size of 100 km in diameter down to diapirs with sizes ranging from 0.3 to 30 km (Kendall and Melosh 2016). Once reduced to sizes below 10 km, hydrodynamic processes continue to reduce the radius of the iron droplets.

2.3.2. *Fragmentation in the magma ocean*

The core fragments dispersed in the Earth's mantle following an impact then undergo further mixing by turbulent entrainment as the molten iron fragments cross the magma ocean and settle deeper into the planet. This second fragmentation forms even smaller iron droplets (with radii of <1 m). This fragmentation takes place in the impact-melted reservoir, implying a medium with very low viscosity (0.01–100 Pa.s) at depths up to the core-mantle boundary (CMB), depending on the impact angle and the size of the impactor (Nakajima et al. 2021).

Given its size and viscosity, the impact-melted reservoir is a potentially highly turbulent medium within which the metallic phase deforms and then fragments into droplets of different sizes, resulting in more or less efficient thermo-chemical exchanges. Fragmentations potentially take place in cascade until the droplets reach a characteristic size, which will depend essentially on the viscosity of the surrounding medium. For likely viscosities of silicate liquids, the diameter of stable droplets is estimated at around 1 cm (Rubie et al. 2003). This led to the formulation of the *iron rain scenario* (Figure 2.2), a scenario widely used to quantify the efficiency of thermo-chemical exchanges between iron and silicates in a magma ocean (Rubie et al. 2003; Ichikawa et al. 2010; Wacheul et al. 2014).

This scenario can be investigated through laboratory experiments. Wacheul et al. (2014) have, for example, used liquid gallium and water–glycerol mixtures as analogues of molten iron and silicates, respectively (Figure 2.7). This type of experiment makes it possible to study the influence of the viscosity ratio between iron and silicates on fragmentation dynamics. The classical scenario of iron rain considers a population of purely spherical drops with a single characteristic radius falling to the bottom of the magma ocean at a single velocity. However, the experiments of Wacheul et al. (2014) showed a variety of stable shapes for the liquid metal drops, a broad distribution of sizes and velocities, and intense internal dynamics within the cloud with the superposition of new fragmentations and drop assembly (Figure 2.7).

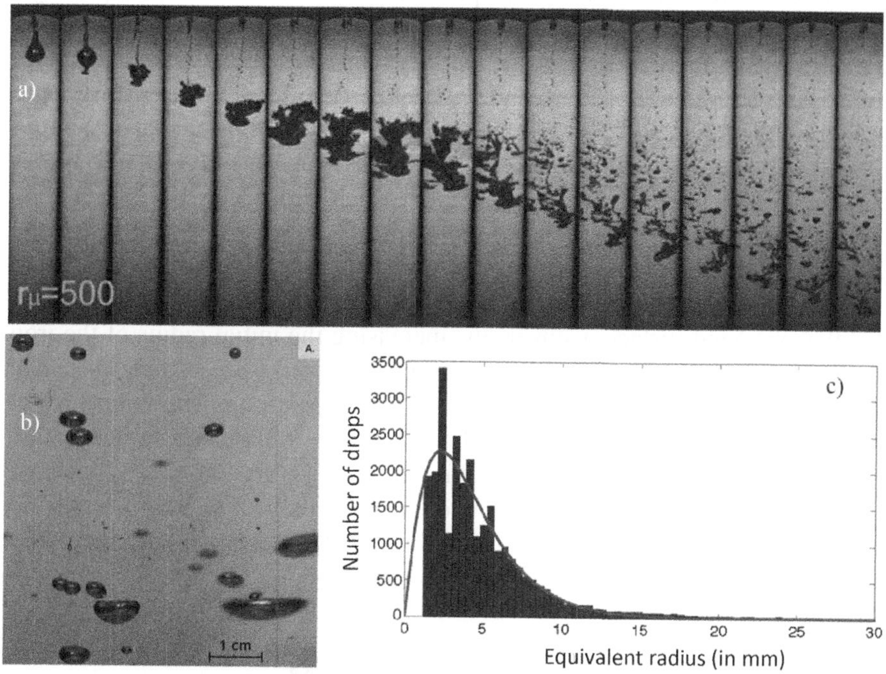

Figure 2.7. *Modeling core/mantle separation using analog laboratory experiments (from Wacheul et al. (2014)). For a color version of this figure, see www.iste.co.uk/monteux/earth1.zip*

COMMENT ON FIGURE 2.7.– *(a) Analogical modeling of the fragmentation of a diapir with an initial radius of 31 mm. We represent the temporal evolution (from left to right) for a case where the ambient fluid is 500 times more viscous than the fragmenting diapir. (b) Liquid gallium droplets of different sizes falling through pure glycerol. (c) Histogram of the equivalent radius of droplets formed by the rupture of a diapir of 30 mm initial radius falling through a mixture of glycerol and water (in blue) and best-fit Gamma distribution (in red) (from Wacheul et al. (2014)).*

Numerical simulations can also be used to characterize the fall dynamics of an initially spherical liquid iron droplet in a molten silicate phase, up to its eventual fragmentation. Numerical work has focused on modeling the fragmentation dynamics of an iron droplet in a magma ocean (Qaddah et al. 2019). By varying the viscosity of the molten silicates between 0.05 and 100 Pa.s, and the initial radius of the iron drop between 1 and 350 mm, these numerical models have made it possible to constrain the morphology, dynamics and stability of an iron drop as a function of the Weber number (ratio between inertial forces and surface tension), the Reynolds number (ratio between inertial forces and viscous forces) and the viscosity ratio between the molten silicates and the liquid iron drop. In particular, this work shows that the maximum radius of the stable drop and the critical Weber number (which determines the range in which drops can be expected to fragment into smaller drops) are monotonically increasing functions of the viscosity of the magma ocean. We now know that increasing the viscosity of the silicate phase has the effect of limiting the oscillations of the metal phase and the exchange surface. Conversely, increasing the initial radius of the iron droplet increases its deformability, thereby increasing its relative exchange surface. Above the critical Weber number, liquid iron fragmentation occurs at a drop distance of between 3.5 and 8 times the initial drop radius in the Weber number range explored. As this distance is very small, in a magma ocean, droplets can be expected to rapidly reach a stable size at which they no longer fragment, and then rapidly equilibrate both chemically and thermally with the surrounding molten mantle.

2.3.3. *Thermochemical evolution of iron droplets*

Numerical models can also track heat exchange between an iron drop and the magma ocean for different initial sizes and viscosities of the surrounding magma. The models developed by Qaddah et al. (2020) show that heat exchange between the two phases occurs mainly in the frontal part of the drop (Figure 2.8). Through a systematic and parametric study, Qaddah et al. (2020) have shown that the thickness of the thermal boundary layer, the depth and time of equilibration, the Nusselt number (the average heat flux at the surface of the drop compared to a purely conductive case) and the volume of the magma ocean affected by heat exchange all follow power laws as a function of the Peclet number (ratio between advective and diffusive heat transfer). Due to the distortions of the drop (Figure 2.8), these scaling laws deviate from classical balances that only consider heat diffusion through a laminar thermal boundary layer. Finally, by considering a temperature-dependent viscosity of the ambient fluid, this study also shows that a low-viscosity layer surrounds the drop, influencing the thermal evolution of non-deformable drops at low Reynolds numbers only, and decreasing the breaking distance for certain breaking modes.

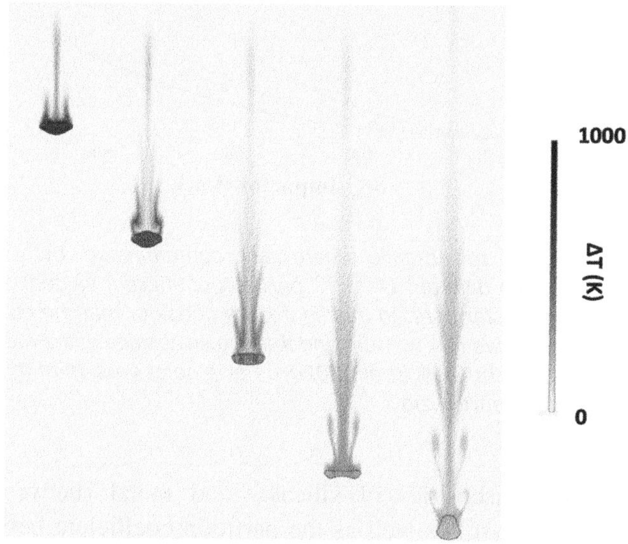

Figure 2.8. *Thermal evolution of a highly deformable metal drop as a function of time. The color scale represents the temperature deviation from the initial temperature. The solid black line represents the surface of the metal drop (from Qaddah et al. (2020)). For a color version of this figure, see www.iste.co.uk/monteux/earth1.zip*

As the chemical and thermal constitutive equations are similar, it is also possible to characterize numerically the iron/silicate chemical equilibration dynamics and derive scaling laws equivalent to those obtained by Qaddah et al. (2020) in order to anticipate the concentration of siderophile elements in the primitive mantle. Clesi et al. (2020) have characterized chemical exchanges during the descent of metal droplets into a magma ocean using numerical models similar to those discussed in the previous section (Figure 2.9).

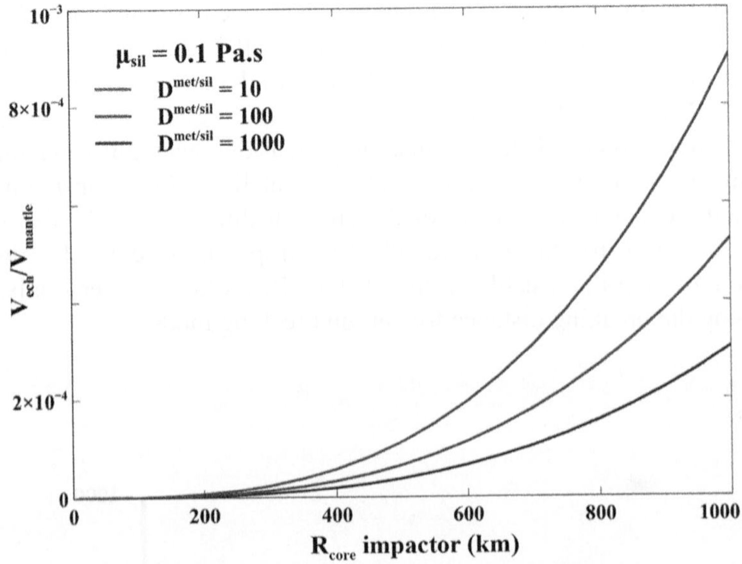

Figure 2.9. *Fraction of the mantle chemically contaminated by a moderately siderophile element (with different $D^{met/sil}$ partition coefficient values) as a function of the radius of the impactor core. In this figure, we consider that the core fragments into iron droplets with radius $R_{Fe} = 1$ cm and that the viscosities of the molten iron and silicates are 0.1 Pa.s. (from Clesi et al. (2020)). For a color version of this figure, see www.iste.co.uk/monteux/earth1.zip*

By varying the viscosities of silicates and metal (between 0.1 and 1000 Pa.s for each phase) as well as the partition coefficient between metal and silicates ($D^{met/sil}$ varying between 1 and 1,000), they derived relationships to quantify the degree of equilibrium between metal and silicate melts, in a regime characterized by low Weber numbers ($We < 10$) and low Reynolds numbers ($10^{-3} < Re < 10^2$). They showed that the main parameters controlling equilibrium for a siderophile element are the viscosity of the silicate phase and the partition coefficient of that element. They applied this parameterization for Ni and Co, two moderately siderophile elements (Figure 2.6), in the context of late terrestrial accretion. Using these models, they were able to constrain the variation of the Ni/Co ratio after a large impact as a function of the viscosity of the magma ocean, for a scenario involving the rain of iron drops in a magma ocean. They show that the volume of a magma ocean potentially contaminated by a moderately siderophile element contained in the metallic phase of an impactor increases with the size of the impactor and decreases as the $D^{met/sil}$ partition coefficient increases (Figure 2.9). While these two results may seem obvious qualitatively, the scaling laws derived by Clesi et al. (2020) nevertheless allow us to quantify these trends. Considering the model of formation of the Earth/Moon system by giant impact (Canup 2004) (see Chapter 3), they showed that such an impact could have a significant influence on the Earth's current Ni/Co ratio. Depending on the radius of Theia's core (the impactor at the origin of the Earth/Moon system) and the viscosity of the magma ocean produced after the impact between proto-Earth and Theia, the formation of the Moon could be responsible for a 0.45–3% increase in the Ni/Co ratio in the present-day silicate mantle for magma ocean viscosities of 0.1–100 Pa.s, respectively.

2.4. Iron/silicate separation by giant diapirism

As the metallic phase separates from the silicate phase within the magma ocean, it accumulates at its base, where there is a major viscosity contrast between the liquid magma ocean and the solid mantle (Figure 2.2). Around 20 orders of magnitude of viscosity separate the magma ocean from the underlying mantle, slowing down iron/silicate separation and promoting the formation of a thick layer of liquid iron that may eventually thermo-chemically equilibrate with the mantle. These large volumes of iron are gravitationally unstable, as they are themselves denser than the underlying solid material composed of silicates (if the planet is differentiated) or an iron/silicate mixture (if the planet is still undifferentiated). As a result, instabilities (known as Rayleigh–Taylor-type instabilities) can form, causing kilometer-sized iron diapirs to fall toward the center of the Earth (Figure 2.2). Honda et al. (1993) have modeled the formation of the Earth's core by considering an iron layer several hundred kilometers thick covering the entirety of a solid, undifferentiated planetary core. Based on their models, they demonstrated that to form the core in <1 billion years, as required by various geophysical and geochemical constraints, the viscosity of the silicate, undifferentiated proto-nucleus would have to be less than 10^{26} Pa.s, which is greater than the current viscosity of the mantle, but much smaller than the viscosity estimated for an undifferentiated planetary core under low-temperature, high-pressure conditions. In particular, this study demonstrated that heating of the undifferentiated planetary core by gravitational energy released during core formation could contribute to the acceleration of the core-forming process. However, it seems unlikely that such a thick layer of iron could form and stabilize before plunging toward the center of the Earth.

Such separation can take place on a global mantle scale, but also on more local scales. A giant impact may melt a large volume of the impacted planet's mantle, forming a magma reservoir whose volume is proportional to the volume of the impactor. The molten material could initially concentrate in a spherical to hemispherical zone below the impact point, in a volume up to two to three times the volume of the impactor (Tonks and Melosh 1993). Mechanical readjustments can then allow this magma reservoir to spread out and cover part or all of the early Earth's surface, depending on its volume. This phenomenon is amplified when several giant impacts follow one another over sufficiently short time intervals to prevent the molten material from crystallizing. Iron collects at the base of these magmatic reservoirs (see previous section) because it is denser than the molten silicates from which it differentiates (Figure 2.10). This iron comes either from the core of the impactor, or from the impacted body itself if iron/silicate separation has not yet taken place in the impact-melted zone. Iron/silicate separation is very rapid when both phases are in the liquid state, but once the iron has collected at the base of the molten zone, this dynamic is slowed down as the plunging speed becomes governed by the viscosity of the surrounding solid silicate phase.

It is possible to numerically model the plunge of an iron diapir collected in the heated zone after a large meteorite impact (Figure 2.10). To study the evolution of such a metallic diapir, analytical models of thermochemical readjustment can be compared with numerical simulations (Monteux et al. 2009). These numerical codes solve the conservation equations for mass, momentum and heat (i.e. the Navier–Stokes equations), taking into account heating by viscous dissipation as well as heating by adiabatic compression/decompression as the iron plunges. These numerical models enable us not only to characterize the dynamics of thermal equilibration, but also the evolution of chemical composition as a function of time. Using these models, Monteux et al. (2009) have shown that the metallic phase sinks with a velocity of the order of the Stokes velocity (velocity proportional to the size of the diapir and the viscosity of the surrounding mantle). Another important point is that the thermal energy released by metal segregation is comparable to the thermal energy dissipated during impact. As detailed in Chapter 3, impact energy is potentially distributed over a large volume of undifferentiated material. The potential energy dissipated during the fall of a large iron diapir, on the other hand, is potentially released in a much smaller volume (the iron diapir and its close surroundings). Significant heating of the metal diapir can occur during the fall, increasing its excess temperature by a factor of 2 or 3 at most. When the viscosity of the metal phase decreases with temperature, the proportion of thermal energy transferred to the surrounding undifferentiated matter increases at the expense of the energy dissipated in the metal phase. This makes it possible to form a proto-nucleus at a temperature close to that of the impact zone, and to induce melting, and therefore additional iron/silicate separation, in the surrounding undifferentiated mantle.

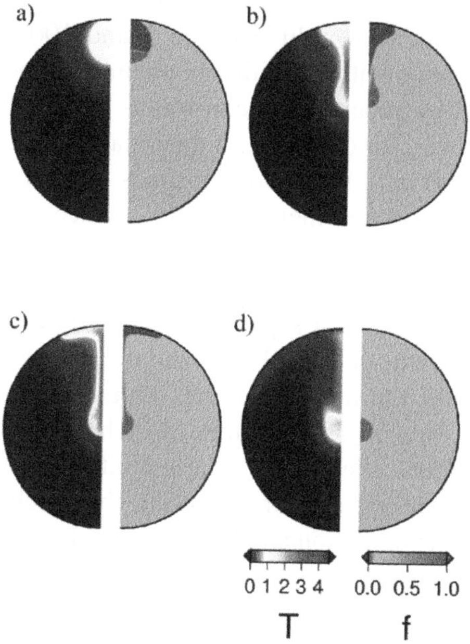

Figure 2.10. *Scaled temperature (left semicircle) and chemical composition (right semicircle) (for iron $f = 1$, for undifferentiated material $f = 0.17$, for silicates $f = 0$) at time $t = 0$ (a), $t = 1,4$ Myr (b), $t = 3.8$ Myr (c) and $t = 546$ Myr (d) (in a case of uniform viscosity, $R = 4,000$ km, $R_{imp} = 600$ km and with a resolution of 200×200 grid points) (from Monteux et al. (2009)). For a color version of this figure, see www.iste.co.uk/monteux/earth1.zip*

This process could be at the origin of a cascading iron/silicate separation in an initially undifferentiated planet. In fact, the potential energy dissipated as heat when a large volume of iron falls leads to additional melting phenomena that help separate more iron, thus dissipating even more potential energy. Šrámek et al. (2010) have developed two-phase models to simulate core-mantle segregation, showing that the first impact that melts the metallic phase near the surface is potentially capable of triggering the entire core-mantle segregation in a runaway phenomenon. They define the threshold of this instability as a function of the size of the impactor and the planet, as well as the planet's initial temperature. Šrámek et al. (2010) have constrained the role of the most important parameters, such as solid and metal viscosities or the depth dependence of gravity.

From a chemical point of view, the formation of the core by large diapirs has consequences for the ability of iron to equilibrate with the surrounding mantle. Metallic diapirs with radii between 1 and 100 m, plunging into a partially molten silicate mantle, equilibrate chemically on timescales in line with those suggested by geochemical constraints (Samuel and Tackley 2008; Samuel 2012). However, for diapirs with much larger volumes (with a radius of between 10 and 100 km), the residence times of the metallic phase in the mantle would be too short and their exchange surface relatively too small to enable efficient chemical exchange and large-scale iron/silicate equilibration in the mantle. The characteristic sizes of iron diapirs, and therefore potentially the impact sizes that enable their separation from the mantle, therefore play a significant role in iron/silicate equilibrium and the composition of siderophile elements in the mantle. On the other hand, large iron diapirs can also carry with them a silicate fraction of the mantle, which then becomes temporarily trapped in the core (Fleck et al. 2018). These molten silicates and the volatile species they contain are capable of triggering thermochemical plumes that would participate in the oxidation and hydration of the upper mantle.

2.5. Giant impact core assemblies

During a giant impact, a large fraction of the core of the impactor can merge with the existing core (Canup 2004). Some, if not all, of the plunging of the metallic phase takes place via diapirism of large volumes of iron. Studies show that giant impacts and the assembly of the two cores have important consequences from a thermal point of view (Arkani-Hamed and Olson 2010), from a chemical point of view (Dahl and Stevenson 2010) and from the point of view of the magnetic history of the impacted object (Reese et al. 2010). These studies focused on the thermomechanical consequences of a giant impact on a differentiated planet. However, none of them considers the merging between the impactor core and the core of the impacted body.

Using numerical models, it is possible to characterize the merging dynamics of the two cores, and in particular the deformation of the pre-existing core and the change in heat flow at the core/mantle boundary

(Figure 2.11) (Monteux et al. 2013; Monteux and Arkani-Hamed 2014). After a giant impact, a significant amount of the impact's kinetic energy is converted into heat within the impacted planet's mantle (Figure 2.11a). The increase in temperature in the mantle decreases the heat flux at the base of the mantle, but also creates thermal stratification beneath the CMB (Arkani-Hamed and Olson 2010). However, as the impactor core falls, viscous heating converting the diapir's potential energy into heat increases the temperature of the plunging core (Figures 2.11(b) and (c)). This second effect increases the temperature in the planet's core and modifies the core's ability to cool by convection (Figure 2.11d). Monteux et al. (2013) have shown that parameters, such as the volume of impactor core retained within the impacted planet and the depth to which this material is buried, govern not only the characteristic assembly time of the two cores but also the intensity of viscous dissipation in the plunging core. Their results also show that significantly increasing the viscosity contrast between the impactor core and the mantle reduces the assembly time of the cores to less than 1,000 years, but does not affect the assembly temperature (Monteux and Arkani-Hamed 2014).

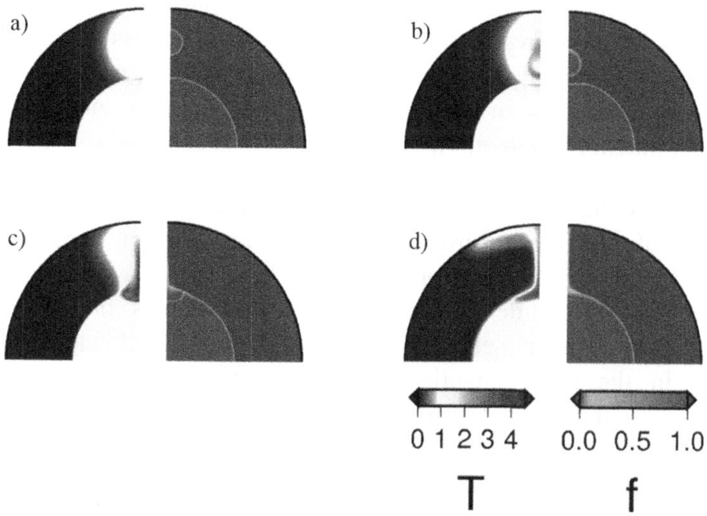

Figure 2.11. *Normalized temperature (left) and composition (right) (for iron* $f = 1$*, for silicates* $f = 0$*) at times t = 0 (a), t = 0.6 Myr (b), t = 6 Myr (c) and t = 38.8 Myr (d) (results obtained with uniform viscosity,* $R_{planete} = 3,400$ *km and* $R_{impactor} = 600$ *km) (Monteux et al. 2013). For a color version of this figure, see www.iste.co.uk/monteux/earth1.zip*

Rayleigh–Taylor instabilities also lead to rapid turbulent mixing of the impactor core with the Earth's mantle. Vortices are formed with characteristic sizes where tungsten (and other siderophile elements) can equilibrate with the magma ocean. However, models developed by Dahl and Stevenson (2010) show that large metal cores (with radii greater than 10 km) plunge into an ocean of silicate magma and do not emulsify entirely by turbulent erosion. This effect, combined with the moderate partition coefficient of Tungsten and the iron/silicate ratio of the mixing material of the order of unity, leads to a moderate equilibration of the core of the impactor at the origin of the Earth–Moon system with the mantle as a whole. This study thus concludes that the ^{182}W excess of the silicate Earth is linked to earlier events rather than to the last giant impact, from which the Moon formed.

2.6. Conclusion and outlook for magnetic fields

Core/mantle separation is a crucial stage in the thermo-chemical evolution of telluric planets, and took place very early in the history of the Solar System, as evidenced by geochemical dating. Despite this early history and the difficulty of accessing deep samples, there are still clues today as to how the planets differentiated, whether in the composition of their more superficial reservoirs or in the dynamics of the planetary interiors (mantle convection, magnetic field, etc.).

During core–mantle differentiation, the metallic phase migrated toward the center of the Earth through processes involving a large range of characteristic sizes and therefore more or less complete thermochemical equilibria. The siderophile elements present in the planets' building material mostly accompanied the metallic phase into the core. Certain elemental ratios measured in the upper mantle give us information on core formation conditions in terms of temperature, pressure and oxygen fugacity. Core/mantle separation probably occurred from a deep magma ocean. The final stages of equilibrium occurred at the base of this magma ocean, at a depth of over 1,000 km, before the iron completed its migration by diapirism. The excess of HSE measured in the rocks of the Earth's mantle proves that an accretion phase occurred after the formation of the core.

The dynamics of core/mantle separation have undoubtedly played a crucial role in the generation of planetary dynamos. To know whether a planetary dynamo can be generated in the core, it is necessary to know the internal thermal state of the planet, and in particular the temperature difference between the core and the surrounding mantle (i.e. the heat flux). Indeed, magnetic field generation requires three ingredients: a liquid metallic fluid with high electrical conductivity, a source of energy provided by thermochemical convection and favorable core dynamics (mainly rotation-induced helicity) (Moffatt 1978; Olson 2007). For bodies with smaller cores (e.g. the Moon), mechanisms other than convection must be invoked (Dwyer et al. 2011; Le Bars et al. 2011). Generally speaking, assuming that the core is essentially composed of iron and that the planet rotates sufficiently fast on itself, the second condition can be characterized by following the evolution of heat flux at the core/mantle boundary. The more vigorous and complex the convection within the core, the greater the probability of having a dynamo (Kutzner and Christensen 2002; Christensen and Aubert 2006). Vigorous convection also translates into significant heat flow at the CMB.

Among the Solar System's telluric planets and moons, Ganymede, Mercury and Earth currently possess a magnetic field. The oldest paleo-magnetic data found on the Earth's surface are dated at 3.45 Ga, so the initiation of geodynamo potentially took place shortly after core formation (Tarduno et al. 2010). Venus has a liquid core but shows no sign of dynamo (Langlais et al. 2010). Mars and probably the Moon have recorded evidence of a transient internal magnetic field in their crust (Dyal et al. 1974; Acuña et al. 1999). Lunar and Martian dynamos are thought to have been initiated early in the history of these objects, only to come to a halt between 4.1 and 3.75 Ga (Acuña et al. 1999; Lillis et al. 2008). Although there is now increasing evidence of these transient dynamos, their chronology and the mechanisms of their initiation and cessation remain major enigmas.

Many models have focused on the influence of giant impacts at the end of planetary accretion on primitive dynamos (Roberts et al. 2009; Arkani-Hamed and Olson 2010; Monteux et al. 2011). Indeed, these giant impacts not only deposit a significant amount of heat in the mantle likely to considerably reduce the cooling efficiency of the core and thus stop the dynamo, but they also potentially bring a significant amount of iron via the core of the impactor which can heat up during its fall and modify the thermal balance of the mantle and core. The temporal frameworks of the magnetic activities of Solar System satellites and planets, and in particular the fact that these activities may have existed during the first few million years of their history, suggest that these dynamos may be linked to the primitive accretion and differentiation stages during planet formation (Stevenson 2003).

2.7. References

Acuña, M., Connerney, J., Ness, N., Lin, R., Mitchell, D., Carlson, C., McFadden, J., Anderson, K., Rème, H., Mazelle, C. et al. (1999). Global distribution of crustal magnetization discovered by the mars global surveyor MAG/ER experiment. *Science*, 284, 790–793.

Agnor, C.B., Canup, R.M., Levison, H.F. (1999). On the character and consequences of large impacts in the late stage of terrestrial planet formation. *Icarus*, 142, 219–237.

Amelin, Y., Krot, A.N., Hutcheon, I.D., Ulyanov, A.A. (2002). Lead isotopic ages of chondrules and calcium-aluminum-rich inclusions. *Science*, 297, 1678–1683.

Arkani-Hamed, J. and Olson, P. (2010). Giant impacts, core stratification, and failure of the Martian dynamo. *Journal of Geophysical Research (Planets)*, 115, 7012.

Beck, P. (2005). Pétrogénèse et métamorphisme de choc des météorites Martiennes. PhD Thesis, Ecole Normale Supérieure de Lyon, Lyon.

Becker, H., Horan, M.F., Walker, R.J., Gao, S., Lorand, J.-P., Rudnick, R.L. (2006). Highly siderophile element composition of the Earth's primitive upper mantle: Constraints from new data on peridotite massifs and xenoliths. *Geochimica et Cosmochimica Acta*, 70, 4528–4550.

Bézos, A., Lorand, J.-P., Humler, E., Gros, M. (2005). Platinum-group element systematics in Mid-Oceanic Ridge basaltic glasses from the Pacific, Atlantic, and Indian Oceans. *Geochimica et Cosmochimica Acta*, 69, 2613–2627.

Bottke, W.F., Walker, R.J., Day, J.M.D., Nesvorny, D., Elkins-Tanton, L. (2010). Stochastic late accretion to Earth, the Moon, and Mars. *Science*, 330(6010), 1527.

Bouhifd, M.A. and Jephcoat, A.P. (2011). Convergence of Ni and Co metal-silicate partition coefficients in the deep magma-ocean and coupled silicon-oxygen solubility in iron melts at high pressures. *Earth and Planetary Science Letters*, 307, 341–348.

Boujibar, A., Andrault, D., Bolfan-Casanova, N., Bouhifd, M.A., Monteux, J. (2015). Cosmochemical fractionation by collisional erosion during the Earth's accretion. *Nature Communications*, 6, 8295.

Bouvier, A., Vervoort, J.D., Patchett, P.J. (2008). The Lu-Hf and Sm-Nd isotopic composition of CHUR: Constraints from unequilibrated chondrites and implications for the bulk composition of terrestrial planets. *Earth and Planetary Science Letters*, 273(1/2), 48–57.

Canup, R.M. (2004). Simulations of a late lunar-forming impact. *Icarus*, 168, 433–456.

Cartier, C., Hammouda, T., Boyet, M., Bouhifd, M.A., Devidal, J.-L. (2014). Redox control of the fractionation of niobium and tantalum during planetary accretion and core formation. *Nature Geoscience*, 7, 573–576.

Chabot, N.L., Draper, D.S., Agee, C.B. (2005). Conditions of core formation in the earth: Constraints from nickel and cobalt partitioning. *Geochemica et Cosmochemica Acta*, 69(8), 2141–2151.

Christensen, U.R. and Aubert, J. (2006). Scaling properties of convection-driven dynamos in rotating spherical shells and application to planetary magnetic fields. *Geophysical Journal International*, 166, 97–114.

Clesi, V., Monteux, J., Qaddah, B., Le Bars, M., Wacheul, J.B., Bouhifd, M.A. (2020). Dynamics of core-mantle separation: Influence of viscosity contrast and metal/silicate partition coefficients on the chemical equilibrium. *Physics of the Earth and Planetary Interiors*, 306, 106547.

Connelly, J.N., Bizzarro, M., Krot, A.N., Nordlund, ÃĚ., Wielandt, D., Ivanova, M.A. (2012). The absolute chronology and thermal processing of solids in the solar protoplanetary disk. *Science*, 338(6107), 651.

Dahl, T.W. and Stevenson, D.J. (2010). Turbulent mixing of metal and silicate during planet accretion and interpretation of the Hf-W chronometer. *Earth and Planetary Science Letters*, 295, 177–186.

Dauphas, N. (2017). The isotopic nature of the Earth's accreting material through time. *Nature*, 541(7638), 521–524.

Deguen, R., Olson, P., Cardin, P. (2011). Experiments on turbulent metal-silicate mixing in a magma ocean. *Earth and Planetary Science Letters*, 310, 303–313.

DeMeo, F.E. and Carry, B. (2014). Solar System evolution from compositional mapping of the asteroid belt. *Nature*, 505(7485), 629–634.

Dwyer, C.A., Stevenson, D.J., Nimmo, F. (2011). A long-lived lunar dynamo driven by continuous mechanical stirring. *Nature*, 479, 212–214.

Dyal, P., Parkin, C.W., Daily, W.D. (1974). Magnetism and the interior of the moon. *Reviews of Geophysics and Space Physics*, 12, 568–591.

Fleck, J.R., Rains, C.L., Weeraratne, D.S., Nguyen, C.T., Brand, D.M., Klein, S.M., McGehee, J.M., Rincon, J.M., Martinez, C., Olson, P.L. (2018). Iron diapirs entrain silicates to the core and initiate thermochemical plumes. *Nature Communications*, 9, 71.

Genda, H., Brasser, R., Mojzsis, S.J. (2017). The terrestrial late veneer from core disruption of a lunar-sized impactor. *Earth and Planetary Science Letters*, 480, 25–32.

Gessmann, C.K. and Rubie, D.C. (2000). The origin of the depletions of V, Cr and Mn in the mantles of the Earth and Moon. *Earth and Planetary Science Letters*, 184(1), 95–107.

Honda, R., Mizutani, H., Yamamoto, T. (1993). Numerical simulation of Earth's core formation. *Journal of Geophysical Research*, 98, 2075–2090.

Ichikawa, H., Labrosse, S., Kurita, K. (2010). Direct numerical simulation of an iron rain in the magma ocean. *Journal of Geophysical Research (Solid Earth)*, 115, 1–12.

Javoy, M., Kaminski, E., Guyot, F., Andrault, D., Sanloup, C., Moreira, M., Labrosse, S., Jambon, A., Agrinier, P., Davaille, A., Jaupart, C. (2010). The chemical composition of the Earth: Enstatite chondrite models. *Earth and Planetary Science Letters*, 293, 259–268.

Kaula, W.M. (1979). Thermal evolution of earth and moon growing by planetesimal impacts. *Journal of Geophysical Research*, 84, 999–1008.

Ke, Y. and Solomatov, V.S. (2009). Coupled core-mantle thermal evolution of early Mars. *Journal of Geophysical Research (Planets)*, 114(13), 1–12.

Kendall, J.D. and Melosh, H.J. (2016). Differentiated planetesimal impacts into a terrestrial magma ocean: Fate of the iron core. *Earth and Planetary Science Letters*, 448, 24–33.

Kleine, T. and Rudge, J.F. (2011). Chronometry of meteorites and the formation of the earth and moon. *Elements*, 7, 41–46.

Kleine, T., Münker, C., Mezger, K., Palme, H. (2002). Rapid accretion and early core formation on asteroids and the terrestrial planets from Hf-W chronometry. *Nature*, 418, 952–955.

Krot, A.N., Amelin, Y., Cassen, P., Meibom, A. (2005). Young chondrules in CB chondrites from a giant impact in the early Solar System. *Nature*, 436(7053), 989–992.

Kutzner, C. and Christensen, U.R. (2002). From stable dipolar towards reversing numerical dynamos. *Physics of the Earth and Planetary Interiors*, 131, 29–45.

Langlais, B., Lesur, V., Purucker, M.E., Connerney, J.E.P., Mandea, M. (2010). Crustal magnetic fields of terrestrial planets. *Space Science Reviews*, 152, 223–249.

Le Bars, M., Wieczorek, M., Karatekin, O., Cébron, D., Laneuville, M. (2011). An impact-driven dynamo for the early Moon. *Nature*, 479, 215–218.

Li, J. and Agee, C.B. (2001). The effect of pressure, temperature, oxygen fugacity and composition on partitioning of nickel and cobalt between liquid Fe-Ni-S alloy and liquid silicate: Implications for the Earth's core formation. *Geochemica et Cosmochemica Acta*, 65(11), 1821–1832.

Lillis, R.J., Frey, H.V., Manga, M. (2008). Rapid decrease in Martian crustal magnetization in the Noachian era: Implications for the dynamo and climate of early Mars. *Geophysical Research Letters*, 35, 14203.

Lorand, J.-P., Luguet, A., Alard, O. (2008). Platinum-group elements: A new set of key tracers for the Earth's interior. *Elements*, 4, 247–253.

Lorand, J.-P., Alard, O., Luguet, A. (2012). Platinum-group elements systematics and petrogenetic processing of the continental upper mantle: A review. *Invited Review Paper, Lithos*, 70, 4528–4550.

Mattern, E., Matas, J., Ricard, Y., Bass, J. (2005). Lower mantle composition and temperature from mineral physics and thermodynamic modelling. *Geophysical Journal International*, 160, 973–990.

Moffatt, H. (1978). *Magnetic Field Generation in Electrically Conducting Fluids*. Cambridge University Press, Cambridge.

Monteux, J. and Arkani-Hamed, J. (2014). Consequences of giant impacts in early Mars: Core merging and Martian dynamo evolution. *Journal of Geophysical Research (Planets)*, 119, 480–505.

Monteux, J., Ricard, Y., Coltice, N., Dubuffet, F., Ulvrova, M. (2009). A model of metal-silicate separation on growing planets. *Earth and Planetary Science Letters*, 287, 353–362.

Monteux, J., Jellinek, A.M., Johnson, C.L. (2011). Why might planets and moons have early dynamos? *Earth and Planetary Science Letters*, 310, 349–359.

Monteux, J., Jellinek, A.M., Johnson, C.L. (2013). Dynamics of core merging after a mega-impact with applications to Mars' early dynamo. *Icarus*, 226, 20–32.

Nakajima, M., Golabek, G.J., Wünnemann, K., Rubie, D.C., Burger, C., Melosh, H.J., Jacobson, S.A., Manske, L., Hull, S.D. (2021). Scaling laws for the geometry of an impact-induced magma ocean. *Earth and Planetary Science Letters*, 568, 116983.

Nimmo, F. and Kleine, T. (2015). Early differentiation and core formation. *Geophysical Monograph Series*, 212, 83–102.

Olson, P. (2007). Overview on core dynamics. In *Treatise on Geophysics*, volume 8, Olson, P. (ed.). Elsevier Science, Amsterdam.

Qaddah, B., Monteux, J., Clesi, V., Bouhifd, M.A., Le Bars, M. (2019). Dynamics and stability of an iron drop falling in a magma ocean. *Physics of the Earth and Planetary Interiors*, 289, 75–89.

Qaddah, B., Monteux, J., Le Bars, M. (2020). Thermal evolution of a metal drop falling in a less dense, more viscous fluid. *Physical Review Fluids*, 5(5), 053801.

Reese, C.C., Orth, C.P., Solomatov, V.S. (2010). Impact origin for the Martian crustal dichotomy: Half emptied or half filled? *Journal of Geophysical Research (Planets)*, 115(14), 5004.

Righter, K. and Drake, M.J. (1999). Effect of water on metal-silicate partitioning of siderophile elements: A high pressure and temperature terrestrial magma ocean and core formation. *Earth and Planetary Science Letters*, 171(3), 383–399.

Roberts, J.H., Lillis, R.J., Manga, M. (2009). Giant impacts on early Mars and the cessation of the Martian dynamo. *Journal of Geophysical Research (Planets)*, 114(13), 4009.

Rubie, D.C., Melosh, H.J., Reid, J.E., Liebske, C., Righter, K. (2003). Mechanisms of metal-silicate equilibration in the terrestrial magma ocean. *Earth and Planetary Science Letters*, 205, 239–255.

Rubie, D.C., Frost, D.J., Mann, U., Asahara, Y., Nimmo, F., Tsuno, K., Kegler, P., Holzheid, A., Palme, H. (2011). Heterogeneous accretion, composition and core-mantle differentiation of the Earth. *Earth and Planetary Science Letters*, 301(1–2), 31–42.

Samuel, H. (2012). A re-evaluation of metal diapir breakup and equilibration in terrestrial magma oceans. *Earth and Planetary Science Letters*, 313, 105–114.

Samuel, H. and Tackley, P.J. (2008). Dynamics of core formation and equilibration by negative diapirism. *Geochemistry, Geophysics, Geosystems*, 9, 6011–6026.

Samuel, H., Tackley, P.J., Evonuk, M. (2010). Heat partitioning in terrestrial planets during core formation by negative diapirism. *Earth and Planetary Science Letters*, 290, 13–19.

Samuel, H., Drilleau, M., Rivoldini, A., Xu, Z., Huang, Q., Garcia, R.F., Lekić, V., Irving, J.C.E., Badro, J., Lognonné, P.H. et al. (2023). Geophysical evidence for an enriched molten silicate layer above Mars's core. *Nature*, 622(7984), 712–717.

Shannon, M.C. and Agee, C.B. (1998). Percolation of core melts at lower mantle conditions. *Science*, 280, 1059.

Šrámek, O., Ricard, Y., Dubuffet, F. (2010). A multiphase model of core formation. *Geophysical Journal International*, 181, 198–220.

Stevenson, D.J. (1981). Models of the earth's core. *Science*, 214, 611–619.

Stevenson, D.J. (2003). Planetary magnetic fields. *Earth and Planetary Science Letters*, 208, 1–2.

Tarduno, J.A., Cottrell, R.D., Watkeys, M.K., Hofmann, A., Doubrovine, P.V., Mamajek, E.E., Liu, D., Sibeck, D.G., Neukirch, L.P., Usui, Y. (2010). Geodynamo, solar wind, and magnetopause 3.4 to 3.45 billion years ago. *Science*, 327, 1238.

Tonks, W.B. and Melosh, H.J. (1992). Core formation by giant impacts. *Icarus*, 100, 326–346.

Tonks, W.B. and Melosh, H.J. (1993). Magma ocean formation due to giant impacts. *Journal of Geophysical Research*, 98, 5319–5333.

Touboul, M., Kleine, T., Bourdon, B., Palme, H., Wieler, R. (2007). Late formation and prolonged differentiation of the Moon inferred from W isotopes in lunar metals. *Nature*, 450, 1206–1209.

Villeneuve, J., Chaussidon, M., Libourel, G. (2009). Homogeneous distribution of ^{26}Al in the Solar System from the Mg isotopic composition of chondrules. *Science*, 325(5943), 985.

Wacheul, J.-B., Le Bars, M., Monteux, J., Aurnou, J.M. (2014). Laboratory experiments on the breakup of liquid metal diapirs. *Earth and Planetary Science Letters*, 403, 236–245.

Walker, R.J. (2009). Highly siderophile elements in the Earth, Moon and Mars: Update and implications for planetary accretion and differentiation. *Chemie der Erde/ Geochemistry*, 69(2), 101–125.

Weber, R.C., Lin, P.-Y., Garnero, E.J., Williams, Q., Lognonné, P. (2011). Seismic detection of the lunar core. *Science*, 331(6015), 309.

Wood, B. (2011). The formation and differentiation of Earth. *Physics Today*, 64(12), 40.

Wood, B.J., Walter, M.J., Wade, J. (2006). Accretion of the Earth and segregation of its core. *Nature*, 441, 825–833.

Yin, Q., Jacobsen, S.B., Yamashita, K., Blichert-Toft, J., Télouk, P., Albarède, F. (2002). A short timescale for terrestrial planet formation from Hf-W chronometry of meteorites. *Nature*, 418, 949–952.

3

Dynamics and Thermal Evolution of the Earth's Early Mantle

Julien MONTEUX and Denis ANDRAULT
LMV, CNRS, Université Clermont Auvergne, IRD, OPGC, Clermont-Ferrand, France

3.1. Introduction

The Earth is a formidable thermal machine that tends to erase traces of the most ancient events. However, geochemical data, the presence of the Moon and certain geophysical observables still preserve the memory of events crucial to the Earth's evolution that took place over 4 billion years ago. Our knowledge of the history of the Earth's interior has benefited from the advent of planetary exploration, numerical calculation and isotope analysis tools, as well as tremendous advances in experimental petrology.

There is a consensus on the importance of the early stages of Earth's formation in the acquisition of its present-day characteristics. However, the modalities of primitive evolution and the processes involved remain widely debated. Comparative planetology and numerical simulations are powerful tools for understanding and reconstructing the phenomena that may have taken place over 4.5 billion years ago. Telluric planets formed rapidly, within the first 100 million years after the formation of the first solids in the Solar System (Kleine et al. 2009), and giant impacts were probably frequent during the final stages of accretion (Canup and Agnor 2000; Jacobson and Morbidelli 2014). The development of geochemical analysis methods has made it possible to constrain the chronology of telluric planet differentiation. The major differentiation events (core formation, magma ocean crystallization) occurred within the first 100 million years (Kleine et al. 2002; Boyet and Carlson 2005).

The Earth's mantle has thus undergone a succession of major events from the formation of the Earth to what we know of it today, thanks in particular to the contribution of seismic tomography. These events, such as giant impacts, core/mantle separation and the solidification of the primitive mantle, have had consequences not only in thermal terms, with major redistributions of primitive heat, and in dynamic terms, with major transitions in flow regimes, but also in chemical terms, with the establishment of more or less complete segregation within the mantle reservoir. These events may also have influenced the way in which the Earth's magnetic field developed. Although the mantle makes up about 85% of the Earth's volume, it remains no less mysterious, and its primitive evolution is still widely debated in the scientific community. Today, most of the mantle is solid and its thermochemical evolution seems rather slow, but this was not always the case.

In this chapter, we detail the energy sources that make the primitive mantle a dynamic and complex reservoir, then attempt to understand the cooling dynamics of this mantle during the major phases of its evolution, and in particular the primitive magma ocean stages and other significant melting stages. Finally, we will characterize the processes that brought the mantle to its current thermochemical properties before considering the influence of the core, atmosphere and Moon on its primitive dynamics.

3.2. Primitive energy sources

The oldest solid samples in the Solar System are the calcium–aluminum-rich inclusions (CAIs) found in undifferentiated meteorites. While these inclusions have been dated at over 4.56 billion years ago (Amelin et al. 2002), geochemical dating shows that iron/silicate separation took place during or just after accretion of the Solar System planets. For the Earth, the calculated ages are highly dependent on the models considered for core formation, and range from 30 to 100 million years after the formation of the CAIs (Kleine et al. 2002). For such segregation to take place so rapidly, it is necessary to involve the silicate phase and/or iron melting processes to facilitate the fall of iron toward the center of the Earth. The segregation process may even have been initiated in the earliest planetary embryos before the final assembly leading to today's planets (Yoshino et al. 2003). This section aims to detail the energy sources that may have contributed to important fusion phenomena during the early evolution of the Earth.

3.2.1. *Short-lived radioactive elements*

The heat released by the decay of short-lived radioactive elements probably played an important role in the temperatures of primitive planetesimals. Indeed, the radioactive decay of certain elements is accompanied by the release of energy in the form of radiation (γ) or the emission of α or β particles, whose diffusion in the surrounding materials releases significant quantities of heat. Isotopic elements (parents) with short half-lives have now disappeared, but an excess of elements (daughters) from this decay can be found in meteorites containing CAIs or chondrules. The study of these now "extinct" radioactivities in meteorites provides important information on the initial formation environment of the first bodies and stars (Meyer and Clayton 2000).

Parent element	$\tau_{1/2}$(Myr)	Daughter element	Heat production per unit mass (W/kg)
^{26}Al	0,7	^{26}Mg	$1{,}5 \times 10^{-7}$
^{238}U	4468,3	^{206}Pb	$9{,}17 \times 10^{-5}$
^{40}K	1248	^{40}Ar	$2{,}97 \times 10^{-5}$
^{232}Th	14010	^{208}Pb	$2{,}56 \times 10^{-5}$
^{60}Fe	1,5	^{60}Ni	$0{,}3 - 1{,}2 \times 10^{-8}$

Table 3.1. *Radioactive isotope pairs that have contributed to the heat balance of the Earth's mantle (according to Lebrun et al. (2013) and McDonough et al. (2020))*

If we consider H the rate of internal heat production by isotopic decay of a planetesimal of radius R, this heating takes precedence over thermal diffusion when $\rho C_p H$ is greater than kT/R^2 with k the planetesimal's thermal conductivity, T its temperature, C_p its heat capacity and ρ its density. As $H \sim T/t$, when $\rho C_p R^2/(kt) >> 1$, the planetesimal warms faster than it cools by diffusion with t the age of the planetesimal. At 1 Myr, and for objects with radii of less than 5 km, diffusion cooling was predominant. At 1 Myr, according to Kokubo and Ida (2000), the largest planetesimals would have reached radii of $\sim 10--100$ km. The heat associated with decay was therefore globally retained. The maximum temperature reached at the center of a planetesimal of radius $R = 30$ km could have exceeded 1,270 K (Schramm et al. 1970; Hutcheon and Hutchison 1989). This would have allowed the largest objects to undergo a possible differentiation associated with fusion phenomena via the decay of short-lived radioactive isotopes, and then dissipate their heat by diffusion. By the time protoplanets have reached large sizes ($\sim 1,000$ km), they have dissipated a significant portion of their heat by diffusion from the radioactive decay of short-lived elements.

Two of these extinct radioactivities may have been involved in the initial thermal state of planetesimals (Table 3.1), the decay of ^{26}Al (mainly localized in the silicate phase of planetesimals) and the decay of ^{60}Fe (mainly localized in the metallic phase of planetesimals), although the importance of this radioactivity is much debated (Goswami and Vanhala 2000). Significant excesses of ^{26}Mg have been found confined in CAIs in carbonaceous meteorites and ordinary chondrites, but also in non-refractory meteorite material (Hutcheon and Hutchison 1989). This presence of excess ^{26}Mg reflects the past presence of ^{26}Al. The ^{26}Al was thus present in significant quantities during the first millions of years of the Solar System.

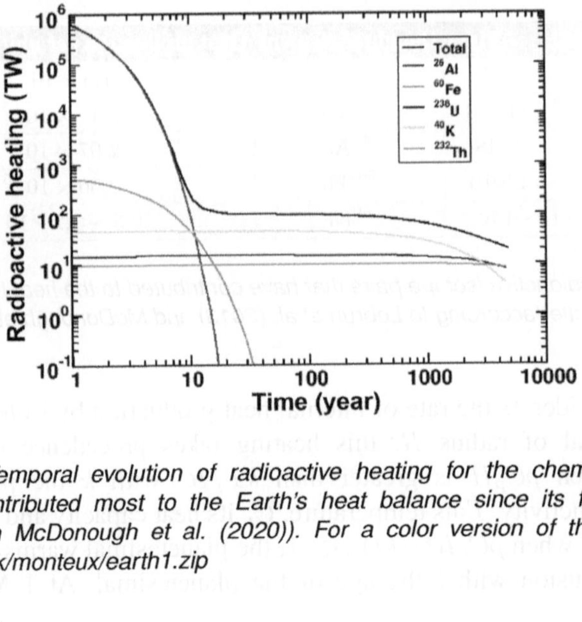

Figure 3.1. *Temporal evolution of radioactive heating for the chemical elements that have contributed most to the Earth's heat balance since its formation over 4.5 Gyr (from McDonough et al. (2020)). For a color version of this figure, see www.iste.co.uk/monteux/earth1.zip*

It is possible to estimate the power dissipated by this heating mode as a function of time (Figure 3.1). This power is a function of the chemical composition considered under initial conditions. McDonough et al. (2020) estimate that during the first 20 million years after the formation of CAIs, ^{60}Fe and ^{26}Al were the elements that contributed most to radioactive heating, dissipating power of up to 10^6 TW. Today, radioactive heating is mainly provided by the decay of ^{238}U, ^{40}K and ^{232}Th, which produce around 20 TW of heating at depth, mainly in the lithosphere.

3.2.2. *Impact heating*

Meteorite impacts played a crucial role in the early stages of Earth's evolution, governing in particular its internal differentiation and the formation of its atmosphere (Abe et al. 2000; Monteux et al. 2009; Ricard et al. 2009; Lebrun et al. 2013). During the final accretion stage, most of the mass of telluric objects accumulates and protoplanets become planets. This stage involved extremely energetic impact phenomena between potentially differentiated planetary embryos. These impacts generated large amounts of local melting and the formation of magma oceans, within which iron and silicates separated to form the core and mantle (Tonks and Melosh 1993). Despite the importance of meteorite impacts in shaping the internal thermochemical structure of the Earth as we know it today, these phenomena are still poorly constrained both qualitatively and quantitatively, and many crucial questions remain unanswered. For example, it is currently unclear how the impactor's kinetic energy is dissipated, and what the thermal state of the protoplanet is after an impact. As meteorite impacts are highly complex and energetic phenomena, a better understanding of them could enable us to better anticipate the various stages of melting and recrystallization of the planet during its accretion, and the final thermal state the planet reaches after repeated impacts.

During the collision, part of the kinetic energy is dissipated within the impacted body in the form of heat, resulting in a local increase in temperature at depth, below the point of impact in a spherical to hemispherical region. The temporal evolution of impactor size probably played an important role in the primitive thermal evolution of planets, since impactor size conditions not only the accretion rate, but also the volume of the heated region following an impact (Pierazzo et al. 1997). Impact velocities are generally between 1 and 10 times the escape velocity of the impacted planet, which corresponds to the velocity required to escape the gravitational attraction of the growing body. Impact velocities are thus essentially a function of the size of the impacted body, and the post-impact energy input increases as accretion proceeds (Figure 3.2). It is possible to estimate to the first order the temperature increase ΔT below the point of impact by considering that:

1) the heated volume is three times the volume of the impactor;

2) the velocity of the impactor is at least equal to the escape velocity of the impacted planet;

3) a fraction γ of the impactor's kinetic energy is dissipated as heat ($\gamma \approx 30\%$), the remainder being used to deform and fragment the impactor and its target.

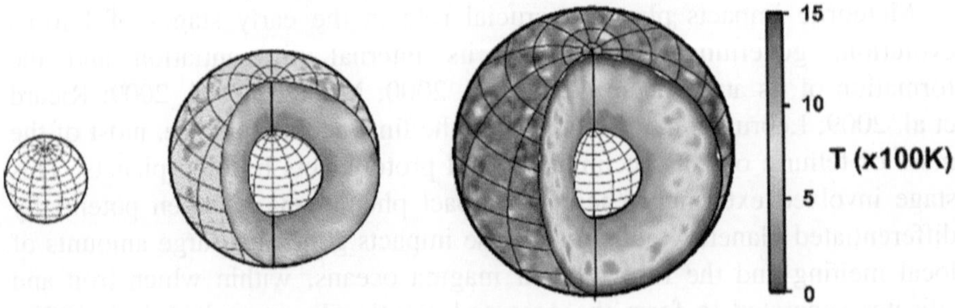

a) b) c)

Figure 3.2. *Representation of the topographic and thermal evolution of a growing planet. For a color version of this figure, see www.iste.co.uk/monteux/earth1.zip*

COMMENT ON FIGURE 3.2.– *In this figure, accretion starts from a 500-km-radius telluric object (a) and culminates in the accretion of a 3,000 km-radius planet. After each impact, the protoplanet enlarges and some of the kinetic energy is dissipated as heat at depth (b). At the end of accretion, as impact velocities are higher, the intensity of impact heating is greater in the outer parts of the planet (c).*

This gives us the following estimate:

$$\Delta T \approx \frac{4\pi\gamma G\rho R^2}{9C_p} \quad [3.1]$$

where G is the gravitational constant, ρ is the density of the impacted body, R is its radius and C_p is its heat capacity. For a body the size of the Moon, the temperature increase is around 100 K, whereas for a body the size of the

Earth, the temperature increase is of the order of 2,000 K (Monteux et al. 2009). Thus, during the final stages of accretion, large volumes of primitive mantle are likely to be heated by impact and melt if the local temperature exceeds the mantle solidus during the planet's growth. Impact energies are thus increasingly likely to melt mantle iron and silicates. This energy can be added to the radioactive decay heat of short-lived elements such as ^{26}Al or ^{60}Fe, leading to significant melting phenomena and the formation of magma oceans.

3.2.3. *Viscous dissipation related to core formation*

Another major source of heat that strongly influenced the primitive thermal state of the Earth's mantle was the formation of its core. If we imagine that the iron in the core was initially distributed uniformly throughout the Earth after accretion, the separation of iron from silicates and the fall of iron to the core resulted in a significant mass transfer, since the iron in the core is denser than the silicates in the mantle. This mass transfer also reduces the potential energy of iron in the mantle, which is converted into heat in the young Earth's interior. Flasar and Birch (1973) estimated that this conversion of potential energy into heat released enough energy to cause the Earth's interior temperature to rise by more than 2,000 K.

The distribution of this heat in the Earth's interior is highly dependent on how the iron migrates to its center (Ke and Solomatov 2009). If the iron plunges along long channels, most of the heat will go into the core. However, if the iron plunges in the form of large drops or diapirs, the potential energy will be dissipated mainly in the mantle. Depending on segregation mechanisms, this will give rise to different thermal profiles, with either a core that is much hotter than the mantle, or a very hot, even molten, mantle. The methods of differentiation, and in particular the characteristic volumes of iron (centimetric droplets or plurikilometric diapirs) brought into play during iron/silicate separation, therefore condition the primitive thermal state of the Earth's mantle. Samuel et al. (2010) have also shown that a large number of small iron diapirs plunging into a magma ocean favors the exchange of iron/silicate heat, leading to a relatively hot core as on Earth. A contrario, a small number of large diapirs has the opposite effect and could correspond to a Mars-like planet. Iron/silicate separation by diapirism tends to leave the deep mantle significantly hotter than its upper part, which could lead to the melting of a significant fraction above the core/mantle boundary.

The viscosity contrast between iron and silicates will also strongly influence the distribution of energy dissipated by viscous friction (Monteux et al. 2009; Qaddah et al. 2020). Monteux et al. (2009) have modeled the evolution of temperature in an iron diapir separated from silicates after heating by a large impact. In particular, they showed that the thermal energy released by metal segregation is lower, but comparable to the thermal energy dissipated during an impact (see previous section). Since impact energy is distributed over a larger volume than the energy dissipated by viscous heating (i.e. the diapir and its immediate surroundings), significant heating of the metal can occur, increasing its excess temperature by a factor of 2 or 3 at most. However, as the viscosity of the plunging iron decreases relative to the viscosity of the material into which it plunges (i.e. the undifferentiated mantle), the proportion of thermal energy transferred to the undifferentiated material increases. This means that heating by viscous dissipation no longer takes place in the low-viscosity plunging metal phase, but in the more viscous surrounding mantle. As a result, the metallic phase will retain the temperature acquired after impact until it reaches the Earth's center, while the undifferentiated mantle may heat up, eventually melting and generating further iron/silicate separation. This phenomenon may even accelerate above a critical impact size for which core/mantle separation may be complete within a planet following a single impact (Ricard et al. 2009).

3.2.4. *Giant impacts and the Moon's formation*

During the final stages of the growth of telluric planets, collisions between protoplanets are hardly exceptional events (Agnor et al. 1999). Giant impacts are now known to be inevitable during the late stages of planetary accretion and may have had major consequences for the late evolution of telluric planets. We know that certain features of the Solar System can be explained by giant impacts. Giant impacts have been invoked to explain the thinness of Mercury's mantle compared to the size of its core (Benz et al. 1988), the retrograde rotation of Venus, the Martian dichotomy and the angular momentum of the planets today. In fact, planetary rotational angular momentum is acquired during accretion by impact with other protoplanets. Impacts can increase or reduce the rotational angular momentum of the impacted planet. The final angular momentum is therefore highly dependent on all previous giant impacts.

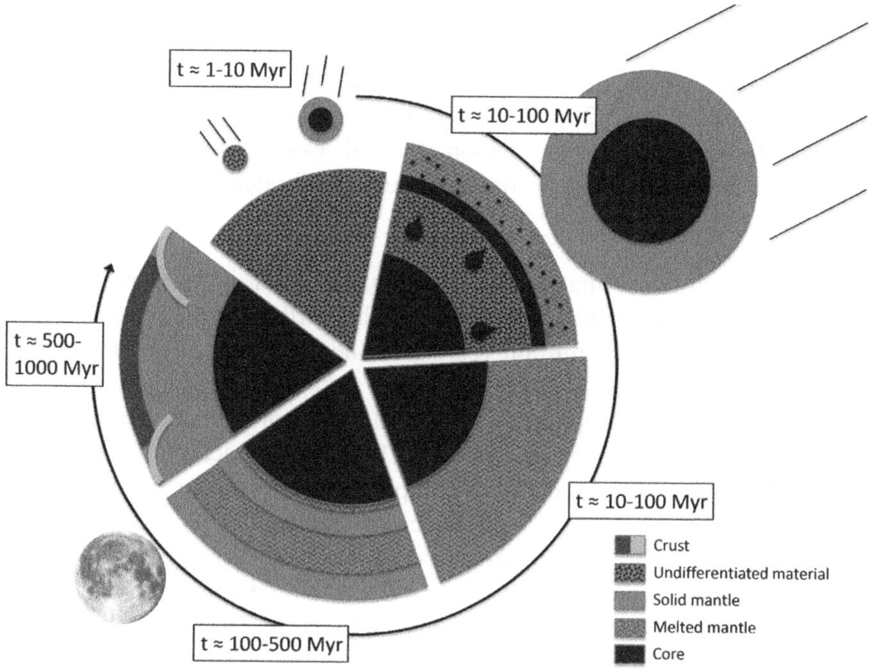

Figure 3.3. *Schematic representation of the evolution of the Earth's internal structure from the end of accretion to modern plate tectonics. For a color version of this figure, see www.iste.co.uk/monteux/earth1.zip*

The Moon's formation is probably due to one or more giant impacts during Earth's early history. The small size of the lunar metallic core relative to the thickness of its mantle, and the fact that the Moon and Earth have similar isotopic compositions, notably in oxygen (Wiechert et al. 2001), support the idea that, early in Earth's history, at a time when it was already differentiated (70–100 Myr after its formation), the Earth was impacted by a large object. The debris field following this impact, essentially made up of mantle-forming terrestrial material and therefore depleted in iron, then re-accreted around the Earth, enabling the formation of the Moon in orbit around the Earth. The cores of the impactor and the Earth mixed to form the Earth's core (Hartmann and Davis 1975). Studies of the angular momentum of the Earth–Moon system by Cameron and Ward (1976) then led to the same conclusion, giving an order of magnitude for the size of the impactor: between 1/3 and 1/2 Earth radius. The impact that formed the Earth–Moon system would have taken place at a velocity of the order of a few $km.s^{-1}$.

This theory has given rise to a great deal of work on modeling this impact, notably using the SPH (*smoothed-particle hydrodynamics*) codes developed by Canup and Asphaug (2001). The similarity of isotopic composition in oxygen (Wiechert et al. 2001) and tungsten (Touboul et al. 2007) suggests two possibilities. Either the Moon formed essentially from terrestrial material, or the lunar and terrestrial mantles equilibrated following the giant impact.

In addition to the formation of the Moon, this giant impact probably had major consequences for the thermal state of the Earth's mantle. The timing of the impact (less than 100 million years after the formation of the first solids) implies that it took place on a very hot planet already significantly affected by previous impacts, radioactive heating and core formation. According to models by Nakajima and Stevenson (2015), the impact that formed the Earth–Moon system resulted in the complete melting of the Earth's mantle and more or less extensive mixing processes depending on the size of the impactor and the Earth's pre-impact rotation speed. Thus, after the giant impact, the mantle would be reduced to a 2,900-km-deep magma ocean with extremely turbulent convective dynamics (Solomatov 2000). However, these results are not supported by geochemical data showing that undifferentiated reservoirs that have not undergone any significant melting episode currently exist within the Earth's mantle, storing chemical elements that are depleted in the rocks sampled and analyzed by geochemists.

3.3. Melting curves in the deep mantle

The solidification of a magma ocean is governed by the melting curves of the primitive mantle (solidus and liquidus), which determine whether the material is solid, partially molten or liquid, and by the temporal evolution of the temperature profile at depth in the reservoir. At the depth at which the temperature profile intersects the liquidus, the first solids are formed and move toward the surface or the bottom of the magma ocean, depending on the density contrast between the crystals and the rest of the magma ocean. So, depending on the chemical composition of the minerals (which governs their density), the melting curves and the temperature profile, different crystallization scenarios can be envisaged: either with a solidification front rising from the bottom of the magma ocean toward the surface, or from the middle of the ocean, thus isolating a basal magma ocean at depth (Labrosse et al. 2007).

Thanks to recent advances in experimental petrology, the melting curves of materials supposed to best represent the mantle are now very well constrained in the laboratory. These experiments require assumptions to be made about the chemical composition of the early mantle. For this, we can either consider a chondritic composition model, which is equivalent to subtracting the composition of the core from the composition of a chondrite, or a pyrolitic model whose theoretical composition is close to that of a peridotite but before the formation of the crust. High-temperature, high-pressure experiments with diamond anvil cells are then used to characterize the temperatures and pressures at which the material begins to melt (solidus) or crystallize (liquidus), up to pressures and temperatures compatible with the conditions prevailing at the core/mantle boundary.

3.3.1. *Mantle solidus profile: the temperature at which the first magmas formed*

To date, there are only a dozen or so articles reporting experimental measurements of the melting temperature of the very deep mantle. This is due to experimental difficulties in reproducing the extreme pressure and temperature conditions prevailing there, and also to difficulties in detecting the melting of a microscopic sample compressed in the diamond anvil cell and heated by an infrared laser. In this section, we will focus on determining the mantle solidus, which corresponds to the temperature at which the mantle begins to melt, and the evolution of the solidus with pressure. The liquidus profile, which corresponds to full mantle melting, is even less well constrained at great depths. Fortunately, liquidus has less impact than solidus on our understanding of the history and properties of the Earth's mantle. The various experimental studies reporting mantle solidus diverge with increasing pressure. The controversy reaches 600–700 K for a pressure of 135 GPa at the core/mantle boundary (Figure 3.4). However, if we develop a detailed analysis of the various studies, the final uncertainty is much smaller. First, there are uncertainties linked to the methods used to determine the sample temperature. It is the thermal emission from the hot core of the sample that is used to measure the temperature (in other words, its color), but the strategies developed for the optical measurement of the emission spectrum and the method of calculating the temperature differ from one study to another. Also, we have limited knowledge of the emissivity factor of each sample, which leads to an uncertainty of several hundred degrees on the temperature.

In Figure 3.4, we plot the melting curves published in previous studies after correcting for some of these effects and using the same calculation assumptions for all. Note first the three highest melting curves, corresponding to peridotite (Fiquet et al. 2010), pyrolite (Pierru et al. 2022) and chondritic-type mantle (Andrault et al. 2011). These curves are highly consistent with each other; peridotite being depleted in minor and fusible elements (e.g. K and Na) compared to the other two, this makes it slightly more refractory. These curves make it possible to constrain the maximum temperature of the core/mantle interface to 4,000 K, since the mantle is mainly solid above this interface.

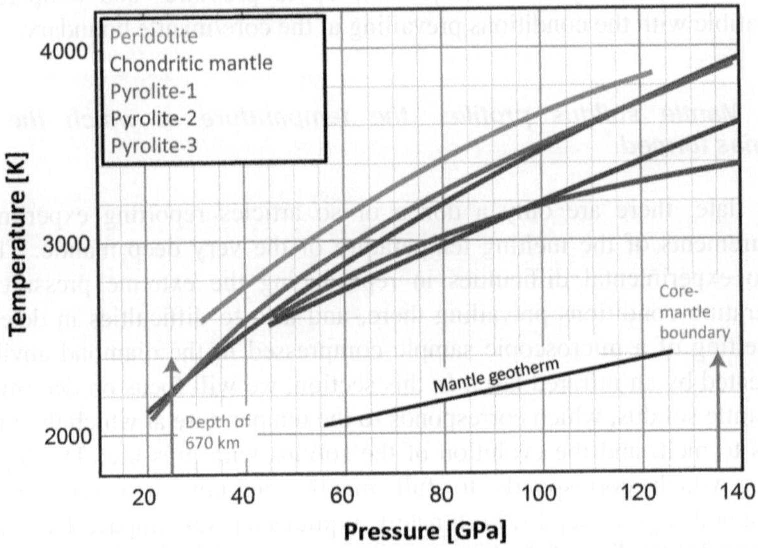

Figure 3.4. *Solidus melting curves for a variety of deep mantle compositions. For a color version of this figure, see www.iste.co.uk/monteux/earth1.zip*

COMMENT ON FIGURE 3.4.– *This figure represents solidus profiles for peridotite, pyrolite and chondritic mantle. Peridotite appears slightly more refractory than the other 2, while there is still significant controversy concerning the melting temperature of pyrolite. The two lowest melting curves, for pyrolite-2 and pyrolite-3, could correspond to the presence of a small amount of water in the samples. We also plot the mantle temperature profile (in black), which is much lower than the mantle melting temperature, whatever the chemical composition model considered. Adapted from Pierru et al. (2022).*

Two other melting curves have been proposed for pyrolitic composition, pyrolite-2 (Nomura et al. 2014) and pyrolite-3 (Kim et al. 2020), which are at significantly lower temperatures than the pyrolite-1 curve (Pierru et al. 2022) discussed earlier. This controversy is necessarily linked to differences in experimental methods, either in the sample itself, in temperature measurement, or in the detection of sample melting. The most likely explanation is the presence of some water in the experimental charges with the lowest melting curves.

3.3.2. *Melting curves for other silicates in the Earth's mantle*

We now plot the melting curves of other silicates that may be present in the deep Earth (Figure 3.5). These are stishovite, the high-pressure phase of quartz (SiO_2), bridgmanite, the most abundant phase in the mantle, with a chemical composition $(Mg_{0.955},Fe_{0.045})(Si_{0.993},Al_{0.007})O_3$ and sometimes simplified to $MgSiO_3$ to facilitate experiments, and a basalt that would have been dragged into the deep mantle by a plunging lithospheric plate. A comparison of these melting profiles with those of the mantle compositions discussed above (Figure 3.4) leads to some interesting observations.

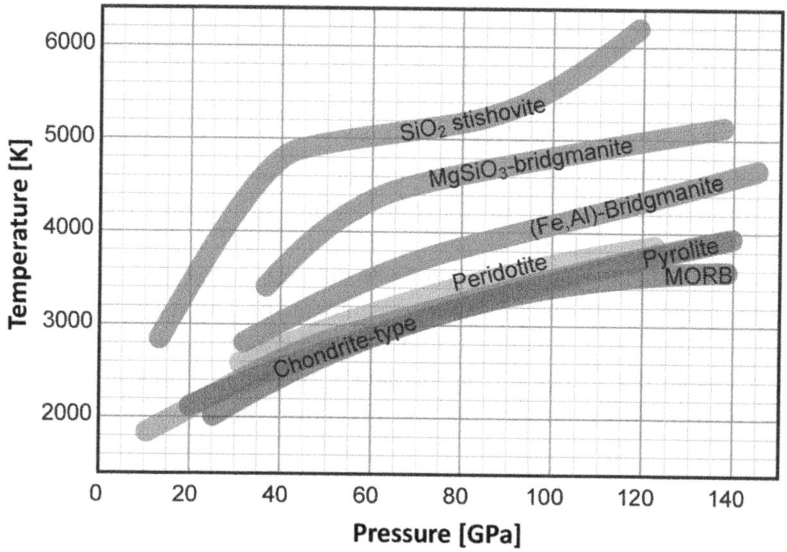

Figure 3.5. *Melting curves for various mantle silicates. For a color version of this figure, see www.iste.co.uk/monteux/earth1.zip*

COMMENT ON FIGURE 3.5.– *The melting temperature decreases as the number of cations in the silicate increases, from SiO_2 stishovite (1 cation only) to peridotitic, pyrolitic and basaltic compositions (reported here as MORB) containing six or more cations. This property is well explained by simple thermodynamic considerations, since the melting of these materials operates in a eutectic-type diagram (see Figure 3.6).*

First, the higher the number of cations in the composition, the lower the solidus melting temperature. This number ranges from 1 to 6 between SiO_2 stishovite and samples of pyrolite, peridotite and basalt. We also note that basalt, which has the highest content of minor fusible elements (Na, K, etc.), has the lowest melting temperature between these different compounds for all pressures representative of the deep mantle. Another observation is that the curvature of the melting profile is more pronounced the lower the number of cations. This change in curvature appears at lower pressures (about 40 GPa) for SiO_2, which contains just one cation (silicon). The change in curvature is visible at around 60 GPa for $MgSiO_3$ bridgmanite, is detectable between 60 and 80 GPa for Fe- and Al-rich natural bridgmanite, and the other melting curves show a very gradual change between 40 and 100 GPa. This curvature denotes a change with pressure in the Clapeyron slope ($dT/dP = \Delta V/\Delta S$). In this context, it corresponds to a decrease in melt volume with increasing pressure.

Previous studies of the atomic structure of disordered materials (liquids and glasses) have shown that pressure induces changes in the coordination of the various cations in liquid silicates. In particular, silicon changes from a tetrahedral environment (four oxygen neighbors) at low pressure to an octahedral environment (six neighbors) at high pressure. The trend observed in Figure 3.5 indicates an earlier and more marked change in the coordination of cations, particularly silicon, when the number of cations in the material is smaller. This also indicates that liquid silicates adopt a very compact atomic structure in the deep mantle. The buoyancy of magmas in the deep mantle then depends above all on their chemical composition.

3.3.3. *Eutectic mantle melting diagram: example of the core/mantle boundary*

There are several approaches to developing a theoretical framework for describing deep mantle melting. One of these is to refine the thermodynamic properties of a set of liquids of simple chemical composition, such as the

$MgSiO_3$, MgO, SiO_2 and CaO poles, and then use classical approximations to combine the properties of these poles with each other to model mantle melting properties. As it happens, the properties of these poles are fairly well described up to extreme pressure and temperature conditions by numerical simulations ab initio (i.e. based on the properties of atoms) or by studies of shock wave propagation in materials (Asimow 2018; de Koker et al. 2013). Another major interest of ab initio calculations in particular is to constrain physical parameters that are very difficult to measure experimentally. This strategy yields important first-order results, but application to the Earth's mantle is unfortunately limited due to the large number of chemical elements in the mantle and the complex chemical properties.

Another approach is to experimentally measure the melting properties of complex mantle materials and to develop thermodynamic models based on these results. If we now focus on the melting properties of the mantle at the core/mantle boundary, which corresponds to about 2,900 km depth and 135 GPa, we observe that the various studies converge toward a mantle melting temperature slightly below 4,000 K (Figure 3.4). Under these conditions, the chemical composition of the magma is not very well constrained experimentally, but there are a few constraints. The most notable come from the melting of peridotite up to around 35 GPa in a multianvil press (Ito et al. 2004) and up to around 80 GPa in the diamond anvil cell (Tateno et al. 2014). These studies give strong constraints on the chemistry of liquids in the deep mantle, even if these experiments did not provide a good constraint on the degree of melting of the material studied. It turns out that it is possible to plot the mantle melting diagram at the lower mantle base (Figure 3.6) using a number of anchor points: (i) the melting of $MgSiO_3$ bridgmanite at a temperature slightly above 5,000 K (see Figure 3.5), (ii) the mantle solidus at about 4,000 K, (iii) even imperfect knowledge of the chemical composition of liquids and (iv) the mantle liquidus temperature (temperature at which the degree of mantle melting reaches 100% (Andrault et al. 2011)). This eutectic melting diagram makes it possible to estimate the mantle melting degree and the chemical composition of liquids above the core/mantle interface for various chemical compositions and at different temperatures (Pierru et al. 2022).

Other implications emerge from this diagram (Figure 3.6): the higher the $MgSiO_3$ content of the mantle, the higher the liquidus temperature; this may affect the solidification mechanisms of the primitive magma ocean. For very high levels of $MgSiO_3$, the disappearance of the feropericlase (Mg,Fe)O and davemoite $CaSiO_3$ phases could increase the mantle melting temperature above 4,000 K; this situation could correspond to the deep mantle if solidification of the magma ocean produced the segregation of an almost pure bridgmanite phase at the core surface. Another observation is that bridgmanite grains are depleted in Fe and Al when they co-exist with a liquid, and this depletion is all the greater the higher the temperature. Finally, in the context of the magma ocean, progressive solidification induces a liquid depleted in $MgSiO_3$ bridgmanite and significantly enriched in FeO, CaO, MgO and Al_2O_3, and therefore in ferropericlase and davemaoite, in relation to the overall chemical composition of the mantle.

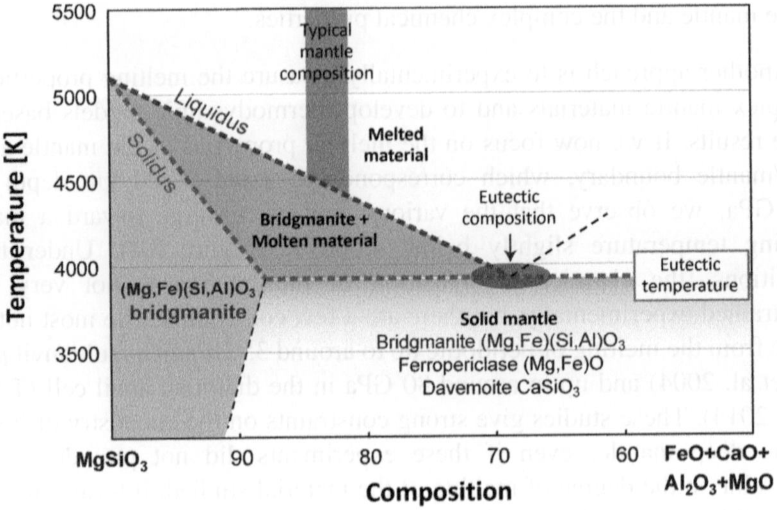

Figure 3.6. *Mantle melting diagram at 135 GPa, the prevailing pressure at the core/mantle boundary. For a color version of this figure, see www.iste.co.uk/monteux/earth1.zip*

COMMENT ON FIGURE 3.6.– *Here, we construct a binary melting diagram between the pure pole consisting of the bridgmanite $MgSiO_3$ and a group of the other major elements of the Earth's mantle, that is, FeO, CaO, Al_2O_3 and MgO. Despite uncertainties about the chemical composition of the deep mantle, we can assume that it should partially melt at temperatures above 4,000 K. The composition of the liquid under these conditions (called eutectic composition) is depleted in $MgSiO_3$ compared to the mantle. Conversely, cooling of a deep magma ocean induces bridgmanite crystallization below 4,500 K. Crystallization will be progressive between 4,500 K and 4,000 K,* *before the complete crystallization of a magma ocean with a chemical composition typical of that of the mantle.*

3.3.4. *Temperature constraints at the core/mantle boundary*

First of all, extrapolation of the mantle geotherm along an adiabatic gradient gives a temperature of around 2,600 K at the core/mantle boundary (Figure 3.4, (Katsura et al. 2010)). But this value does not take into account the strong temperature gradient expected in the thermal boundary layer just above the core. A significantly higher value, between 2,800 and 3,400 K, is obtained by performing a joint inversion of the seismic profiles and elastic (compressional and shear) properties of the minerals (Matas et al. 2007). Unfortunately, this type of modeling is not very sensitive to temperature changes in the thermal boundary layer.

Seismic observations in this very deep region show a generally solid mantle, even in contact with the core. This implies that the temperature is below the mantle solidus, that is, below around 4,000 K. Otherwise, the mantle would be partially molten wherever it is in contact with the core. However, seismic evidence points to small regions of very low seismic velocity, the ULVZs (*ultra-low velocity zones*), which seem to correspond to partially melted zones (Yu and Garnero 2018). It is therefore likely that the temperature at the core/mantle interface is not much lower than the mantle solidus. In this case, a local excess of impurities (such as Na and K in a basalt), hydrogen or CO_2 could produce a lowering of the melting temperature and the formation of these ULVZs.

However, the nature of the chemical heterogeneities in ULVZs remains highly debated. If it were a basalt, the temperature would have to be higher than 3,600 K to cause partial melting; if it were the presence of hydrogen (water, in other words), the temperature would have to be higher than around 3,400 K (Figure 3.4). These various constraints imply a temperature of between 3,400 and 4,000 K at the core/mantle boundary. Note that, on the other hand, the melting properties of iron alloys typical of the outer core suggest a temperature of around 4,000 K (Anzellini et al. 2013).

3.4. The mantle during the magma ocean stage

During the final stages of accretion, the Earth's mantle probably underwent several episodes of more or less complete melting known as the "magma ocean stage", during which the temperature profile was significantly higher than the melting curves discussed above. Indeed, the formation of the Earth's core took place very early in its evolution (between 30 and 50 Myr) (Kleine et al. 2002), requiring a high melting rate to accommodate rapid iron/silicate separation. Moreover, geochemical data show that a crust may have formed very early (between 30 and 200 Myr), as would be expected at the surface of a magma ocean (Boyet and Carlson 2005). While there is consensus on the existence of these magma ocean episodes, many uncertainties remain as to their lifespan, the evolution of the fraction of molten material during cooling, the degree of chemical fractionation during solidification and the couplings with the underlying metallic core and early atmosphere.

3.4.1. *A vigorous convection system*

The dynamics of a magma ocean thicker than 1,000 km is extremely turbulent (Solomatov 2007). In order to quantify the vigor of convection in a reservoir potentially involving the entire thickness of the mantle, it is possible to calculate the Rayleigh number, which is the ratio between the characteristic time of heat transfer by conduction and the characteristic time of heat transfer by advection:

$$Ra = \frac{\alpha \rho g \Delta T H^3}{\eta \kappa} \qquad [3.2]$$

where α is the coefficient of thermal expansion of the mantle, ρ is its density, g is the Earth's gravity, ΔT is the temperature difference between the top and bottom of the magma ocean, H is the thickness of the magma ocean, η is its viscosity and κ is its thermal diffusivity. To convect, a fluid reservoir must have an Ra greater than a critical value Ra_c. For the earth's mantle, $Ra_c \approx 1,000$. Below this critical value, cooling takes place solely by conduction, whereas for $Ra > Ra_c$, heat transfer takes place by advection and conduction: convection. The Rayleigh number in today's mantle is estimated at $10^6 - 10^8$, indicating that it must be convecting. Considering that the magma ocean covers the entire mantle and that the viscosity of a liquid magma is between 10^{-2} and 10^2 Pa.s, the Rayleigh numbers obtained are between 10^{20} and 10^{30}, indicating that convection is very vigorous in this environment. Characteristic flow sizes are thus all the smaller with small sub-metric eddies. To characterize such dynamics, several approaches are possible.

3.4.2. *Magma ocean modeling*

To understand the dynamics of systems with very high Rayleigh numbers, we can use analog laboratory models involving a tank filled with gas (He or SF_6, for example) heated from below and cooled from above (Roche 2020). Coupled with theoretical analytical models (Siggia 1994), these experiments make it possible to constrain the cooling of a highly turbulent fluid and, in particular, to propose relationships between the heat flux observed at the reservoir surface (characterized by the Nusselt number, Nu) and the strength of convection (characterized by the Rayleigh number, Ra) (see Chapter 5 of this volume for more details). Different convection regimes can thus be highlighted, ranging from turbulent (with $Nu \propto Ra^{1/3}$) to ultra-turbulent (with $Nu \propto Ra^{2/7}$), in which magma oceans are probably found (Solomatov 2007). Other analogical models can also be used to quantify the stability of a layer of crystals in a reservoir under laminar and turbulent convection (Solomatov and Stevenson 1993).

Another modeling approach is numerical. It involves solving the conservation equations for momentum, mass and heat using a numerical code (see Chapter 1 of this volume for details). Numerous models of magma ocean evolution have been developed in one dimension (Solomatov and Stevenson 1993; Abe 1997; Monteux et al. 2016) and have the following advantages: (1) they can characterize short time scales and their spatial resolution can reach that of the characteristic lengths of a medium in turbulent convection; (2) 1D models can model extreme viscosity contrasts such as those existing between solid and liquid fraction in a magma ocean; (3) constraints derived from recent experimental petrological data, such as melting curves (solidus and liquidus) or parameters whose value may change with depth, such as density or viscosity, can easily be implemented in these models. However, while 1D models calculate average temperature profiles over the entire depth of the reservoir, they cannot characterize lateral heterogeneities in temperature or density. It is therefore impossible for these models to track the initiation of solid convection, which can take place even before the end of magma ocean crystallization, or to highlight overturn phases generated by density or temperature gradients, which is possible with numerical modeling tools (2D or 3D; (Maurice et al. 2017)). Moreover, since 1D models are unable to calculate the velocity field, they have to make strong assumptions about the thicknesses of boundary layers that set up in zones where the flow velocity decreases sharply (at the bottom or surface of the magma ocean, or close to crystallized and stable zones).

3.4.3. *Adiabatic profiles*

In a system such as the earth's mantle, pressure is hydrostatic and increases with depth. If this medium is at equilibrium, temperature increases with pressure and therefore with depth, following an adiabatic profile (or adiabat). Thermal heterogeneities, linked for example to the core being very hot, cause instabilities in this system that can make the temperature gradient greater than the adiabatic gradient, resulting in thermal convection. When this convection is vigorous, the temperature gradient is slightly greater than the adiabatic gradient. In order to estimate the temperature profile in a magma ocean, it is therefore possible to calculate the adiabatic temperature profile according to the relationship:

$$\frac{dT}{dP} = \frac{\alpha T}{\rho C_p} \qquad [3.3]$$

where T is the temperature, P is the pressure, α is the coefficient of thermal expansion, C_p is the heat capacity and ρ is the density in the mantle. Thus, assuming a potential temperature T_p, which is the temperature of the adiabat at the Earth's surface at $P = 0$, it is possible to calculate first-order temperature profiles in the mantle (Figure 3.7).

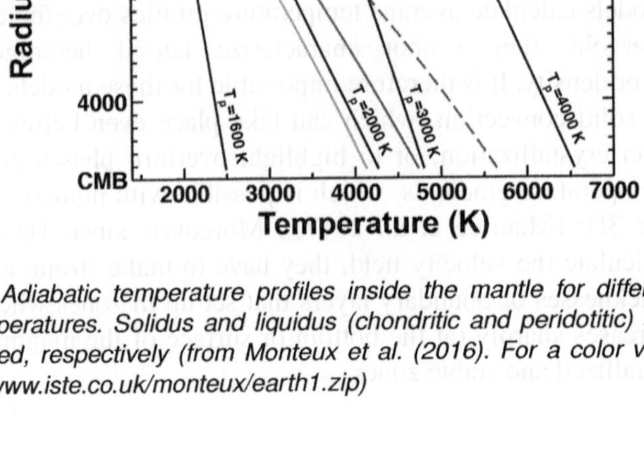

Figure 3.7. *Adiabatic temperature profiles inside the mantle for different potential surface temperatures. Solidus and liquidus (chondritic and peridotitic) are shown in green and red, respectively (from Monteux et al. (2016). For a color version of this figure, see www.iste.co.uk/monteux/earth1.zip)*

Currently, petrological data allow us to estimate that the Earth's deep mantle follows an adiabat with a potential temperature of around 1,600 K. However, this temperature profile was probably much higher just after the giant impact that formed the Earth–Moon system. If we assume that the entire mantle was melted, the characteristic potential temperature of the post-impact mantle was necessarily higher than 3,000 K (Figure 3.7).

3.4.4. *Cooling dynamics*

A giant impact, coupled with an increase in temperature, probably melted the entire Earth's mantle. In this section, we will attempt to describe the cooling of a magma ocean that would include the entire Earth's mantle. After the impact, the core was potentially very hot, while the Earth's surface was at an equilibrium temperature that could exceed 3,000 K depending on the characteristics of the primitive atmosphere (Lebrun et al. 2013). Rapidly after this giant impact, the magma ocean began to convect in order to evacuate its internal heat via the Earth's surface. The surface temperature was much lower, while the core was still hot. After rapid cooling from the surface, the temperature dropped sharply and fell below the liquidus temperature at the level of contact with the atmosphere. Here, a very thin and unstable crust formed, similar to that at the surface of a lava lake. Two extremely thin thermal boundary layers developed (1) at the base of the mantle, where silicate material contrasts with metallic material, and (2) at the surface of the magma ocean at the boundary between the atmosphere and the mantle. These boundary layers are associated with a discontinuity in the velocity field, and thus with heat transfer by conduction only. The temperature profile inside the magma ocean rapidly adjusted to an adiabatic profile (Figure 3.8). Cooling induces a gradual lowering of this adiabatic internal profile, with an increasingly lower potential temperature.

The mantle continues to cool globally until the temperature profile crosses the liquidus (Figure 3.8). In a chondritic mantle model, this crossover occurs at the core/mantle boundary and magma ocean solidification will therefore develop from the bottom of the mantle toward the surface (Monteux et al. 2016). However, if the crossover occurs in the middle of the mantle, two solidification fronts can be set up: one from the middle of the mantle toward the surface, and one from the middle toward the core/mantle boundary. In this case, a basal magma ocean can be isolated beneath the solid mantle (Labrosse et al. 2007).

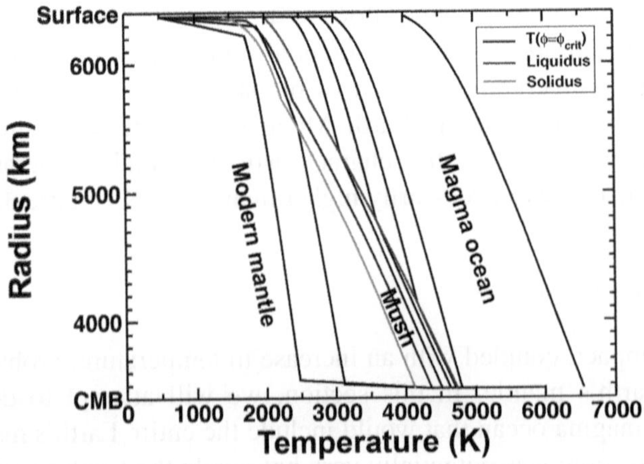

Figure 3.8. *Temporal evolution of temperature profiles inside the Earth's mantle from global melting (rightmost profile) to the current profile (leftmost profile). For a color version of this figure, see www.iste.co.uk/monteux/earth1.zip*

COMMENT ON FIGURE 3.8.– *Chondritic solidus and liquidus are shown in green and red, respectively. The temperature profile in blue represents the profile at 40% melt fraction (i.e. 60% crystals). At this profile, the mantle transitions from the magma ocean stage to the mush stage, which is much more viscous (from Monteux et al. (2016, 2020)).*

During its early history, the magma ocean underwent a sudden slowdown in cooling when its temperature profile reached values corresponding to 40–50% melt fraction (i.e. a temperature profile between liquidus and solidus corresponding to 50–60% crystallization) (Figure 3.8). This slowdown could have occurred less than 1 million years after the impact that formed the Earth/Moon system, depending on the viscosity of the completely liquid magma (Monteux et al. 2016). At this point, the viscosity of the material increases sharply as crystals begin to interact mechanically with each other and prevent fluid deformation. The magma/mineral mixture called "mush" has a rheological behavior close to that of a solid. The fraction of solid material increases until a threshold separating the turbulent regime from the viscous solid-state regime is reached (Abe 1997). As soon as this threshold is reached, the cooling efficiency of the primitive mantle becomes limited. This has important consequences not only for the thermal state of the deep mantle, but also for the cooling capacity of the core, and thus for the Earth's magnetic history. During its early history, the magma ocean underwent a sudden slowdown in cooling when its temperature profile reached values corresponding to 40–50% melt fraction (i.e. a temperature profile between liquidus and solidus corresponding to 50–60% crystallization) (Figure 3.8). This slowdown could have occurred less than 1 million years after the impact that created the Earth/Moon system, depending on the viscosity of the completely liquid magma (Monteux et al. 2016). At this point, the viscosity of the material rises sharply as crystals begin to interact mechanically with each other and prevent fluid deformation. The magma/mineral mixture called "mush" has a rheological behavior close to that of a solid. The fraction of solid material increases until a threshold separating the turbulent regime from the viscous solid-state regime is reached (Abe 1997). As soon as this threshold is reached, the cooling efficiency of the primitive mantle becomes limited. This has important consequences not only for the thermal state of the deep mantle, but also for the cooling capacity of the core, and thus for the Earth's magnetic history.

A second, much slower phase of cooling will then take place once the entire mantle is at around 40% melt fraction. During this phase, the viscosity of the solid phase becomes critical, as it governs not only the rate of cooling but also the depths at which the last pockets of molten material are maintained.

3.5. From magma oceans to present-day mantle dynamics

During the magma ocean stage, the primitive terrestrial mantle underwent rapid global cooling until its molten fraction decreased to a critical value ϕ_c ($\sim$40%) associated with a significant increase in its viscosity (Solomatov 2000; Monteux et al. 2016). This was followed by a slow cooling stage that resulted in almost complete crystallization of the silicate mantle (Solomatov 2000). This second cooling stage could have lasted hundreds of Myr, or even a few Gyr, as its dynamics was governed by the rheology of the slowly deforming solid mantle, unlike the first, which was governed by the viscosity of the magma (Solomatov 2000; Ulvrová et al. 2012; Monteux et al. 2016, 2020).

The depth at which complete crystallization is reached is probably governed by (1) the solidus temperature which controls the depth and temperature of solidification, (2) the thickness of the upper thermal boundary layer where convective heat from the deep mantle is transferred to the surface by conduction, (3) the temperature contrast on both sides of the thermal boundary layer which controls the efficiency of heat evacuation and (4) the primitive atmosphere which could slow mantle cooling (Lebrun et al. 2013; Salvador et al. 2017). In the partially molten regime, the viscosity of the solid mantle is a key parameter, governing mantle dynamics and the formation of thermal boundary layers. Today, mantle viscosity is difficult to constrain, particularly in the lower mantle (Cížková et al. 2012), and the value of mantle viscosity prior to the onset of major differentiation events is even more hypothetical. Constraining the value of this physical quantity is of prime importance, as deep solid mantle viscosity probably governs the initiation of major geological features, such as plate tectonics and mantle plumes (Sleep et al. 2014; Foley et al. 2014).

3.5.1. *Melt fraction rates and viscosity models*

The fraction of molten material ϕ_{melt} plays a fundamental role in the viscosity of the primitive mantle (η) and therefore in its cooling dynamics. As the melt fraction decreases in a cooling magma, there is a transition from a flow behavior more typical of the liquid phase to one more characteristic of the solid phase. As the melt fraction decreases, a threshold is reached at which the solid fraction is sufficiently large that solid grains can no longer enter suspension in the fluid. When this transition is reached, a major rheological transition occurs, with viscosity increasing by several orders of magnitude (Figure 3.9).

3.5.1.1. $\phi_{melt} = 1$

Above the liquidus where $\phi_{melt} = 1$ (= 100%), the mantle material behaves like a liquid. In this case, the viscosity of the liquid phase η_l depends essentially on its composition, temperature T and pressure P. Karki and Stixrude (2010) developed a computational and experimental approach to constrain the viscosity of the $MgSiO_3$ liquid phase at pressures and temperatures compatible with those of the earth's mantle. They obtained the following relationship:

$$\eta = \eta_l = \exp(-7.75 + 0.005P(\text{GPa}) - 0.00015P(\text{GPa})^2 + \frac{5,000 + 135P(\text{GPa}) + 0.23P(\text{GPa})^2}{T - 1,000}) \quad [3.4]$$

According to estimates by Karki and Stixrude (2010), the viscosity of the liquid silicate phase could vary between 0.001 and 1 Pa.s. Viscosity increases with pressure and decreases with temperature.

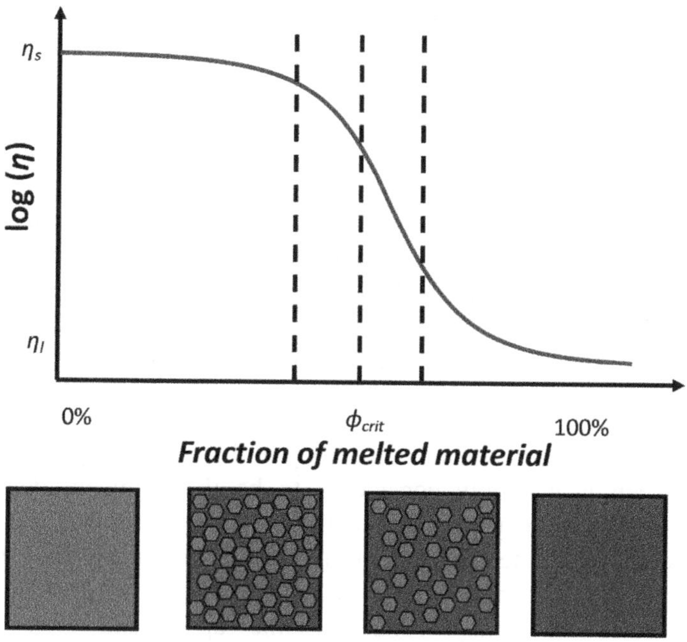

Figure 3.9. *Schematic representation of changes in the viscosity of a partially melted mantle as a function of the solid fraction. For a color version of this figure, see www.iste.co.uk/monteux/earth1.zip*

3.5.1.2. $\phi_{melt} = 0$

The viscosity of the solid fraction in the deep mantle of the early Earth is a key parameter, governing its cooling efficiency during the phase following the magma ocean stage. This parameter is difficult to constrain, whether for the present-day deep mantle (Čížková et al. 2012) or for the primitive mantle, even though the viscosity of a "bridgmanite" deep mantle is increasingly documented (Boioli et al. 2017; Reali et al. 2019). The two main deformation mechanisms associated with mantle convection are diffusion creep and dislocation creep. In the case of diffusion creep, deformation is linked to atomic diffusion of crystalline defects, and this diffusion is thermally activated. In the case of dislocation creep, deformation is accommodated by sliding along dislocations that are lines of imperfections in the crystalline matrix. Experimental and theoretical models for calculating viscosity agree on the following expression, common to both mechanisms:

$$\eta = \eta_s = \frac{1}{A}\exp(\frac{E + PV}{RT}) \qquad [3.5]$$

where P is the pressure, R is the gas constant (= 8.314 J K^{-1} mol^{-1}) and T is the absolute temperature. A is the viscosity pre-factor, E is the activation energy and V is the activation volume for the appropriate creep mechanism (diffusion or dislocation). These three parameters (A, E and V) can be constrained in the present-day deep mantle by lithospheric subduction velocities (Cížková et al. 2012).

Due to the absence of seismic anisotropy in the present-day lower mantle, diffusion creep is generally considered to be the dominant deformation mechanism in the deep mantle (Boioli et al. 2017; Reali et al. 2019). Early in Earth's history, mantle temperatures were higher, and ionic diffusion (hence diffusion creep) should have been even more important in the solid mantle fraction (Frost and Ashby 1982) for the upper and lower mantle. Consequently, we can assume that deformation of the interconnected solid phase occurs solely by diffusion creep. Interestingly, grain size plays a role in the value of the viscosity pre-factor A (Monteux et al. 2020).

3.5.1.3. $0 < \phi_{melt} < 1$

To calculate the viscosity of a partially melted mantle, we need to develop a formalism that takes into account the properties of both the solid and liquid phases. Here, we detail the formalism proposed by Lebrun et al. (2013). During the final stages of magma ocean solidification, the fraction of solid material increases until it reaches a threshold ($\phi_{melt} = \phi_{crit}$), which separates the "liquid" regime from the "solid" regime (Solomatov 2007). This critical fraction of molten material can be estimated by experiments involving migration of molten material into aggregates of solid material (Riley and Kohlstedt 1991) or by tri-axial compression experiments at high temperatures (Scott and Kohlstedt 2006). ϕ_{crit} depends on numerous parameters (pressure, temperature, chemical composition, grain size, etc.). Scott and Kohlstedt (2006) estimated that ϕ_{crit} was between 0.25 and 0.3 for a peridotite at a pressure of 300 MPa and a temperature of around 1,500 K. For a deep magma ocean, ϕ_{crit} could reach 0.4 (Solomatov 2000).

For $\phi_{crit} < \phi_{melt} < 1$, the crystal fraction remains relatively low and the crystal-magma mixture behaves like a liquid. Thus, the viscosity of the partially molten material is a function of the viscosity of the molten magma η_l (Roscoe 1952):

$$\eta = \frac{\eta_l}{(1 - (\frac{1-\phi_{melt}}{1-\phi_{crit}}))^{2.5}} \quad [3.6]$$

In the context of a partially molten mantle with a high crystallization rate ($0<\phi_{melt} < \phi_{crit}$), the crystal–magma mixture tends to behave more like a solid, due to the strong interaction between the crystals. In this case, the viscosity of the material depends on the viscosity of the solid phase η_s and can be expressed as follows:

$$\eta = \eta_s \exp(-26\phi_{melt}) \quad [3.7]$$

During cooling, once the melt fraction threshold is reached in most of the mantle, the cooling efficiency of the primitive mantle decreases considerably, even though the mantle may remain partially molten at certain depths (Monteux et al. 2016, 2020).

3.5.2. *Cooling dynamics*

We previously discussed that the cooling of a completely molten mantle was a rapid process that slowed down abruptly when a significant part of the mantle fell below 40–50% melt fraction. This critical point is reached on time scales between 10^3 and 10^5 years (Abe 1997; Monteux et al. 2016). When this melt fraction is reached in most of the mantle, it passes from the magma ocean stage to the mush stage (a mixture of solid and liquid) (Figure 3.8).

Despite a substantial increase in viscosity following significant solidification, the Rayleigh numbers achieved in the mantle remain very high and convection remains vigorous. The high Rayleigh numbers involved in cooling make it difficult to characterize the dynamics of the mantle as a whole in 2D or 3D. So 1D models may once again be relevant for this kind of approach (Abe 1997; Lebrun et al. 2013; Monteux et al. 2016). Temperature profiles will tend to follow adiabatic profiles. As illustrated in Figure 3.10, most of the mantle and particularly the lower mantle will solidify in less than 90 Myr, although the zone in contact with the core-mantle boundary (i.e. the basal thermal boundary layer) may reach a temperature above the mantle solidus ($\sim$ 4,000 K). In the upper mantle, adiabats may be locally higher than the solidus and partially melted zones may persist.

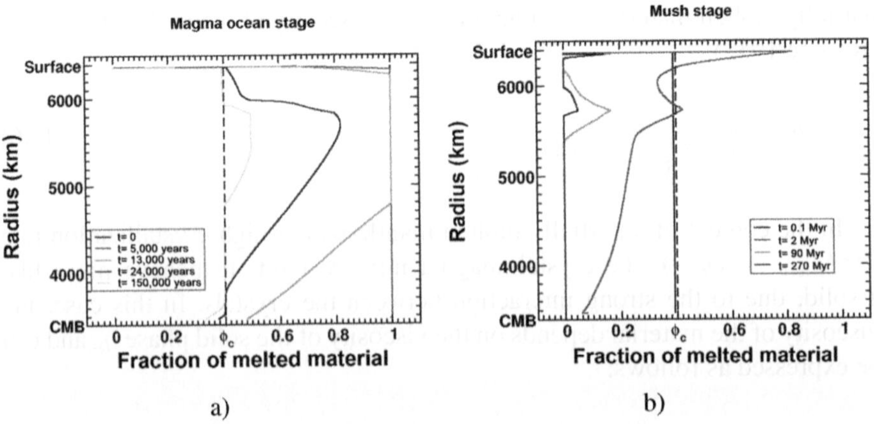

Figure 3.10. *Temporal evolution of the fraction of material in the molten mantle as a function of radius from the center of the Earth during the magma ocean stage (left) and the mush stage (right). From Monteux et al. (2016, 2020). For a color version of this figure, see www.iste.co.uk/monteux/earth1.zip*

Recent models show that the lifespan and depths of the last molten zones in the upper mantle are strongly dependent on the parameters considered to calculate the viscosity of the upper mantle (see equation [3.5]). Depending on the viscosity profile in the mantle, it is thus possible to maintain two partially melted layers at depth: (1) a layer located at depths between 500 and 800 km for 40–400 Myr, and (2) a more superficial layer (30–50 km deep) for more than 500 Myr (Monteux et al. 2020). By combining these time scales obtained by numerical models with geological observations, it is thus possible to provide constraints on the viscosity of the deep primitive mantle, on the parameters used to calculate this viscosity (see equation [3.5]) and on certain parameters such as grain size (Monteux et al. 2020).

3.5.3. *Magma ocean-solid mantle coupling*

During the cooling and solidification of a magma ocean, the juxtaposition of materials with a high melt fraction and a solid material can have significant consequences for the cooling dynamics of the primitive mantle. Such consequences may be recorded in the signatures (geological, geochemical or magnetic) observed on the surface today.

The presence of a magma ocean may thus have played an important role in determining how convection was initiated in the solid phase of the early mantle. Maurice et al. (2017) have developed numerical models that characterize the influence of a magma ocean on the initiation of solid convection. In particular, they have shown that if the magma ocean is present for a million years or more, solid convection can start even before crystallization of the primitive mantle is complete. This would have the major consequence of prematurely erasing any chemical heterogeneities generated during chemical differentiation of the Earth's mantle. Moreover, the presence of a magma ocean in contact with the solid mantle favors latent heat exchanges between the solid and liquid phases, and thus phase changes. These phase changes can lead to material transfers that affect solid mantle convection during crystallization of molten zones. As a result, a solid mantle sandwiched between two molten layers could experience a significant increase in its ability to dissipate heat (Agrusta et al. 2020).

Recent experimental results have shown that the upper mantle solidus is at a lower temperature than expected for a chondritic composition (Andrault et al. 2018). According to this study, such a solidus associated with a warmer mantle would allow the presence of a deep and persistent molten layer in the Archean mantle. The progressive solidification of this molten layer could have strengthened the mechanical coupling between lithosphere and asthenosphere. Such a change could explain the transition from surface dynamics dominated by a *stagnant lid* to modern plate tectonics with deep subductions. The intersection between the mantle solidus and an adiabatic temperature profile with a potential temperature corresponding to a fraction of $\phi_C \sim 40\%$ could constrain the bottom depth of the remaining partially molten layer. In the case of solidification of a partially melted mantle from the core/mantle boundary to the surface, the depth of the last remaining partially melted layer should therefore be less than 300 km as proposed by Solomatov (2007). Complete crystallization of the mantle should occur less than $\sim$1 Gyr (Sleep et al. 2014) after the impact at the origin of the Earth–Moon system.

Finally, the presence of a shallow molten layer during the Hadean and its final crystallization around the Hadean–Eoarchean boundary could prevent the appearance of high orogenic formations due to the rapid isostatic compensation of any relief created and the development of large-scale tectonic faults and shear zones (Monteux et al. 2020). The absence of relief would in turn favor an Earth where most of its surface would be covered by shallow water. Major faults and shear zones could represent pathways for liquid water to penetrate to the lower levels of the crust and, in turn, induce intense hydrothermal activity. Moreover, this weak layer at the depth of the lower crust could eventually isolate the crust from the underlying mantle.

3.5.4. *Geochemical consequences*

The magma oceans were initially the site of more or less complete equilibria and segregations between the metallic phase of the impactors and the earth's molten mantle, while the silicate fraction of the impactor was more simply added to the target mantle (Rubie et al. 2003). This first major phase of differentiation at the origin of the formation of the Earth's core was followed by a second phase of chemical segregation during crystallization of the primitive mantle (Boyet and Carlson 2005). This differentiation is responsible for the separation between compatible elements that will readily join the solid fraction (e.g. Mg and Cr) and incompatible elements that will tend to remain in the liquid fraction (e.g. Al, Na and Fe). As cooling progresses, the compositions of the solid and liquid phases will diverge from the "primitive" mantle composition. Thus, this second major chemical differentiation in the primitive mantle was strongly influenced by the crystallization mechanisms of the magma ocean. The exchanges that took place during this period strongly influenced the geochemical signature of the Earth's interior used today to constrain differentiation processes in terms of characteristic times, P/T conditions or redox conditions.

The buoyancy of the molten phase is in fact governed by the ability of heavier elements such as iron to integrate the solid or liquid phase (Funamori and Sato 2010). This capacity can be estimated in the laboratory under P/T conditions characteristic of the deep mantle by calculating the partition coefficient (D_i) of an element i between a silicate solid phase (bridgmanite or olivine polymorphs) and the liquid phase. While partition coefficients are now increasingly well constrained over a wide range of pressures/temperatures, the geodynamic and geochemical implications of early mantle differentiation are still debated. For example, work on iron partitioning at high pressures can lead to different interpretations: Nomura et al. (2011) suggest a liquid very rich in iron and denser than the mantle, while Andrault et al. (2012) suggest the upwelling of the liquid phase toward the surface, while forming a solid residue depleted in incompatible elements at depth.

The difference in density between solid and liquid can lead to a separation between the two phases (Caracas et al. 2019) via crystal sedimentation processes in the magma ocean. This can produce major chemical segregation between deeper and shallower geochemical reservoirs, whether the buoyancy difference is positive or negative. The geodynamic implications will be very different in each case. The vigor of convection will potentially prevent sedimentation of the solid phase, hence it is crucial to model the global dynamics of this segregation. The ascent or descent of the liquid phase in a partially molten mantle could contribute to volcanism at the surface or to the establishment of a magma ocean at the base of the mantle.

3.6. External influences

3.6.1. *Relationship between the primitive mantle and the core*

The Earth's magnetic field is the result of rapid movements of iron in the liquid part of the Earth's core (Jones 2015). Understanding the thermal state of the Earth's core and its evolution over time therefore also enables us to better understand the evolution of the Earth's magnetic field, and thus why some planets have one and others do not (Buffett 2002). The temporal framework of the Earth's magnetic activity, and in particular the fact that this activity may have existed during the first few million years of its history, suggests that this dynamo is probably linked to the stages of accretion and primitive thermal evolution of the Earth (Stevenson 2003).

As the magma ocean cools, significant heat exchange takes place between the core and the mantle. These exchanges occur by conduction through the thermal boundary layers located on either side of the core/mantle boundary. When convection is very vigorous, as is the case in the liquid metal core and magma ocean, these layers are very thin and the conductive heat flow ($\propto k\Delta T/\delta$) is very high. As a result, the magma ocean stage can promote core cooling if the core is initially hotter than the mantle at its contact. This characteristic can have major consequences for the way the Earth's magnetic field is set up, and for the sources of buoyancy needed to maintain it for several billion years (Buffett 2002).

The way in which the early mantle cooled played a key role in the crystallization of the solid inner core. Indeed, the state of the core (liquid or solid) is governed by the efficiency of heat transport in the mantle and by the thermal energy accumulated in the core at the start of its history. Unless we consider a primitive core at very high temperature, the thermal energy stored in the core is probably insufficient to generate a magnetic field for more than 4 billion years. We therefore invoke the crystallization of the inner core around 1 billion years ago to provide the chemical buoyancy needed to maintain a magnetic field until the present day (Labrosse et al. 2001). However, if the primordial energy of the core was rapidly depleted during the magma ocean stage, it would be necessary to invoke other sources of buoyancy to maintain a magnetic field from its appearance more than 4 Gyr to the present day (Andrault et al. 2016; Badro et al. 2016).

3.6.2. *Influence of the primitive atmosphere*

Similarly, the thickness of the thermal boundary layer beneath the Earth's surface and its early atmosphere played a key role in the thermal evolution of the early mantle. Indeed, these two features undoubtedly governed the efficiency with which heat from the magma ocean was removed. While the thickness of the thermal boundary layer is a function of the internal dynamics of the magma ocean (Monteux et al. 2016), the ability of the atmosphere to retain heat or not is essentially a function of its structure and composition, notably in CO_2 and H_2O. The work of Lebrun et al. (2013) has shown that the presence of a primitive atmosphere can not only delay the formation of the underlying thermal boundary layer but also that an increase in water content can significantly prolong the magma ocean stage on Earth.

As a planet's water content is a function of its formation distance from the Sun, there could be a critical distance beyond which magma ocean stages are short-lived, as was the case on Earth (Hamano et al. 2013). Below this critical distance, with lower water content, magma ocean stages could last close to 100 million years, as on Venus. This results in a more efficient outgassing and planetary interiors less rich in volatile elements.

3.6.3. *The Moon's influence*

Shortly after the impact that created the Earth–Moon system, the Moon was completely re-accreted from the debris of the collision (Kokubo et al. 2000). At that time, the Moon was much closer to the Earth than it is today (Chen and Nimmo 2016), which may have had important consequences for the interaction between the two objects. The transfer of rotational energy into heat by viscous dissipation probably produced more than 1,000 TW during the first tens of millions of years after the Moon's formation impact, which may have had a significant influence on the cooling dynamics of the partially molten Earth. Even if this period of intense dissipation was brief, it could have had a major impact on the crystallization sequence. While the recent orbital and rotational states of the Earth–Moon system are well documented (Williams 2000), little is known about the early interactions within the Earth–Moon system. In particular, the influence of the early Moon on thermal evolution and melting within the early Earth is still poorly understood.

Tidal interactions have played a crucial role in the evolution of the Earth–Moon system, affecting both the heat balance of the two bodies and their rotational-orbital history (Chen and Nimmo 2016). The evolution of the lunar orbit is highly dependent on the way in which tidal energy was dissipated on the primitive Earth, and is therefore strongly coupled to its thermal evolution. On the modern Earth, tidal dissipation is dominated by oceanic dissipation processes (Ray et al. 2001). During the Hadean era, the molten fraction in the Earth's silicate mantle is expected to have a major impact on tidal dissipation processes and thus on the entire evolution of the Earth–Moon system. In turn, the energy dissipated in the Earth during the Hadean may affect its thermal structure, convective dynamics and favor additional melting episodes, leading to a tidally driven thermal regime comparable to that observed on Io, the Moon or Jupiter (Kervazo et al. 2021). Even after the first stage of magma ocean crystallization, a significant melt fraction can be maintained in the Hadean mantle (Monteux et al. 2020) and can control both heat generation by tidal dissipation and heat transfer by melt extraction.

3.7. Conclusion

The first few million years played a fundamental role in the acquisition of the current structure and dynamics of the Earth's mantle. In addition to providing material during accretion, impacts bring energy that is dissipated deep into the core of the impacted body. Together with the heat dissipated during core formation and radiogenic heat, impacts have had a strong influence on the thermal evolution of the mantle. As a result, massive melting episodes have punctuated the history of the young Earth, especially during giant impacts such as the one that formed the Earth/Moon system.

These melting episodes have strongly influenced the chemical structure of the Earth's interior, triggering major episodes of segregation between more or less dense and/or more or less compatible elements. As a result, there may still be reservoirs in the deep mantle today that have not undergone melting and have not been mixed by mantle convection. These reservoirs could explain certain chemical anomalies measured today in samples taken from the Earth's surface.

The dynamics of the Earth's mantle also needs to be understood in the context of the inner Solar System, where other planets such as Mercury, Venus and Mars show certain similarities with the Earth: all these objects have undergone relatively rapid accretion from meteorites of varying size, they are differentiated into a metallic core and a silicate mantle, their mantle composition is probably close to that of the Earth, and they have (Mercury) or had (Mars, Venus) a magnetic field generated by convective motions within their metallic core early in their history. While Venus experienced one or more global resurfacing events in the past, perhaps associated with episodes of subduction, only the Earth today shows signs of vigorous mantle convection. The absence of a current magnetic field and of plate tectonics on Venus is a surprising feature since the Earth and Venus are very similar in size. We can therefore legitimately wonder about the uniqueness of the internal dynamics of the Earth's mantle and its influence on the presence of a thick atmosphere and the development of life on its surface.

An ingredient not to be overlooked is the presence of a massive moon rotating around the Earth. While the Moon protected us from meteorite impacts early in Earth's history, the giant impact that triggered the formation of the Earth/Moon system and the ensuing massive melting episode are potentially the major elements that explain Earth's uniqueness. Moreover, the Moon's presence probably contributed significantly to thermal dynamics through tidal effects, and its presence today may be helping to maintain a magnetic field generated in the liquid core (Andrault et al. 2016). What if the search for extraterrestrial life involved the search for exomoons?

3.8. References

Abe, Y. (1997). Thermal and chemical evolution of the terrestrial magma ocean. *Physics of the Earth and Planetary Interiors*, 100, 27–39.

Abe, Y., Ohtani, E., Okuchi, T., Righter, K., Drake, M. (2000). Water in the early Earth. In *Origin of the Earth and Moon*, Canup, R.M. and Righter, K. (eds). University of Arizona Press/Lunar and Planetary Institute, Tucson/Houston.

Agnor, C.B., Canup, R.M., Levison, H.F. (1999). On the character and consequences of large impacts in the late stage of terrestrial planet formation. *Icarus*, 142, 219–237.

Agrusta, R., Morison, A., Labrosse, S., Deguen, R., Alboussiére, T., Tackley, P.J., Dubuffet, F. (2020). Mantle convection interacting with magma oceans. *Geophysical Journal International*, 220(3), 1878–1892.

Amelin, Y., Krot, A.N., Hutcheon, I.D., Ulyanov, A.A. (2002). Lead isotopic ages of chondrules and calcium-aluminum-rich inclusions. *Science*, 297, 1678–1683.

Andrault, D., Bolfan-Casanova, N., Nigro, G.L., Bouhifd, M.A., Garbarino, G., Mezouar, M. (2011). Solidus and liquidus profiles of chondritic mantle: Implication for melting of the Earth across its history. *Earth and Planetary Science Letters*, 304, 251–259.

Andrault, D., Petigirard, S., Lo Nigro, G., Devidal, J.-L., Veronesi, G., Garbarino, G., Mezouar, M. (2012). Solid-liquid iron partitioning in Earth's deep mantle. *Nature*, 487(7407), 354.

Andrault, D., Monteux, J., Le Bars, M., Samuel, H. (2016). The deep Earth may not be cooling down. *Earth and Planetary Science Letters*, 443, 195–203.

Andrault, D., Pesce, G., Manthilake, G., Monteux, J., Bolfan-Casanova, N., Chantel, J., Novella, D., Guignot, N., King, A., Itié, J.-P. et al. (2018). Deep and persistent melt layer in the Archaean mantle. *Nature Geoscience*, 11(2), 139–143.

Anzellini, S., Dewaele, A., Mezouar, M., Loubeyre, P., Morard, G. (2013). Melting of iron at Earth's inner core boundary based on fast X-ray diffraction. *Science*, 340(6131), 464–466.

Asimow, P.D. (2018). Melts under extreme conditions from shock experiments. In *Magmas Under Pressure: Advances in High-Pressure Experiments on Structure and Properties of Melts*, Kono, Y. and Sanloup, C. (eds). Elsevier, Amsterdam.

Badro, J., Siebert, J., Nimmo, F. (2016). An early geodynamo driven by exsolution of mantle components from Earth's core. *Nature*, 536(7616), 326–328.

Benz, W., Slattery, W.L., Cameron, A.G.W. (1988). Collisional stripping of Mercury's mantle. *Icarus*, 74, 516–528.

Boioli, F., Carrez, P., Cordier, P., Devincre, B., Gouriet, K., Hirel, P., Kraych, A., Ritterbex, S. (2017). Pure climb creep mechanism drives flow in Earth's lower mantle. *Science Advances*, 3(3), e1601958.

Boyet, M. and Carlson, R.W. (2005). ^{142}Nd evidence for early (>4.53 Ga) global differentiation of the silicate earth. *Science*, 309, 576–581.

Buffett, B.A. (2002). Estimates of heat flow in the deep mantle based on the power requirements for the geodynamo. *Geophys. Res. Lett.*, 29(12), 1–4.

Cameron, A.G.W. and Ward, W.R. (1976). The origin of the Moon. *Lun. Planet. Inst. Conf. Abs.*, 7, 120–122.

Canup, R.M. (2004). Simulations of a late lunar-forming impact. *Icarus*, 168, 433–456.

Canup, R.M. and Agnor, C.B. (2000). Accretion of the terrestrial planets and the Earth-Moon system. In *Origin of the Earth and Moon*, Canup, R.M. and Righter, K. (eds). University of Arizona Press, Tucson.

Canup, R.M. and Asphaug, E. (2001). Origin of the Moon in a giant impact near the end of the Earth's formation. *Nature*, 412, 708–712.

Caracas, R., Hirose, K., Nomura, R., Ballmer, M.D. (2019). Melt-crystal density crossover in a deep magma ocean. *Earth and Planetary Science Letters*, 516, 202–211.

Chen, E.M.A. and Nimmo, F. (2016). Tidal dissipation in the lunar magma ocean and its effect on the early evolution of the Earth-Moon system. *Icarus*, 275, 132–142.

Cížková, H., van den Berg, A.P., Spakman, W., Matyska, C. (2012). The viscosity of Earth's lower mantle inferred from sinking speed of subducted lithosphere. *Physics of the Earth and Planetary Interiors*, 200, 56–62.

Fiquet, G., Auzende, A.L., Siebert, J., Corgne, A., Bureau, H., Ozawa, H., Garbarino, G. (2010). Melting of peridotite to 140 gigapascals. *Science*, 329, 1516.

Flasar, F.M. and Birch, F. (1973). Energetics of core formation: A correction. *J. Geophys. Res.*, 78, 6101–6103.

Foley, B.J., Bercovici, D., Elkins-Tanton, L.T. (2014). Initiation of plate tectonics from post-magma ocean thermochemical convection. *Journal of Geophysical Research (Solid Earth)*, 119(11), 8538–8561.

Frost, H.J. and Ashby, M. (1982). *Deformation-Mechanism Maps: The Plasticity and Creep of Metals and Ceramics*. Pergamon, Oxford.

Funamori, N. and Sato, T. (2010). Density contrast between silicate melts and crystals in the deep mantle: An integrated view based on static-compression data. *Earth and Planetary Science Letters*, 295, 435–440.

Goswami, J.N. and Vanhala, H.A.T. (2000). Extinct radionuclides and the origin of the Solar System. In *Protostars and Planets IV*, Mannings, V., Russell, S.S., Boss, A.P. (eds). University of Arizona Press, Tucson.

Hamano, K., Abe, Y., Genda, H. (2013). Emergence of two types of terrestrial planet on solidification of magma ocean. *Nature*, 497(7451), 607–610.

Hartmann, W.K. and Davis, D.R. (1975). Satellite-sized planetesimals and lunar origin. *Icarus*, 24, 504–514.

Hutcheon, I.D. and Hutchison, R. (1989). Evidence from the Semarkona ordinary chondrite for Al-26 heating of small planets. *Nat*, 337, 238–241.

Ito, E., Kubo, A., Katsura, T., Walter, M.J. (2004). Melting experiments of mantle materials under lower mantle conditions with implications for magma ocean differentiation. *Physics of the Earth and Planetary Interiors*, 143, 397–406.

Jacobson, S.A. and Morbidelli, A. (2014). Lunar and terrestrial planet formation in the Grand Tack scenario. *Royal Society of London Philosophical Transactions Series A*, 372, 174.

Jones, C. (2015). Thermal and compositional convection in the outer core. In *Treatise of Geophysics*, volume 8, Olson, P (ed.). Elsevier, Amsterdam.

Karki, B.B. and Stixrude, L.P. (2010). Viscosity of $MgSiO_3$ liquid at Earth's mantle conditions: Implications for an early magma ocean. *Science*, 328(5979), 740.

Katsura, T., Yoneda, A., Yamazaki, D., Yoshino, T., Ito, E., Suetsugu, D., Bina, C., Inoue, T., Wiens, D., Jellinek, M. (2010). Adiabatic temperature profile in the mantle. *Physics of the Earth and Planetary Interiors*, 183(1/2), 212–218.

Ke, Y. and Solomatov, V.S. (2009). Coupled core-mantle thermal evolution of early Mars. *Journal of Geophysical Research (Planets)*, 114(13), 1–12.

Kervazo, M., Tobie, G., Choblet, G., Dumoulin, C., Běhounková, M. (2021). Solid tides in Io's partially molten interior. Contribution of bulk dissipation. *Astronomy and Astrophysics*, 650, A72.

Kim, T., Ko, B., Greenberg, E., Prakapenka, V., Shim, S.-H., Lee, Y. (2020). Low melting temperature of anhydrous mantle materials at the core-mantle boundary. *Geophys. Res. Lett.*, 47(20), e89345.

Kleine, T., Münker, C., Mezger, K., Palme, H. (2002). Rapid accretion and early core formation on asteroids and the terrestrial planets from Hf-W chronometry. *Nature*, 418, 952–955.

Kleine, T., Touboul, M., Bourdon, B., Nimmo, F., Mezger, K., Palme, H., Jacobsen, S.B., Yin, Q.-Z., Halliday, A.N. (2009). Hf-W chronology of the accretion and early evolution of asteroids and terrestrial planets. *Geochimica et Cosmochimica Acta*, 73, 5150–5188.

de Koker, N., Karki, B.B., Stixrude, L. (2013). Thermodynamics of the MgO-SiO_2 liquid system in Earth's lowermost mantle from first principles. *Earth and Planetary Science Letters*, 361, 58–63.

Kokubo, E. and Ida, S. (2000). Formation of protoplanets from planetesimals in the solar nebula. *Icarus*, 143, 15–27.

Kokubo, E., Ida, S., Makino, J. (2000). Evolution of a circumterrestrial disk and formation of a single moon. *Icarus*, 148(2), 419–436.

Labrosse, S., Poirier, J.-P., Le Mouël, J.-L. (2001). The age of the inner core. *Earth and Planetary Science Letters*, 190, 111–123.

Labrosse, S., Hernlund, J.W., Coltice, N. (2007). A crystallizing dense magma ocean at the base of the Earth's mantle. *Nature*, 450, 866–869.

Lebrun, T., Massol, H., Chassefière, E., Davaille, A., Marcq, E., Sarda, P., Leblanc, F., Brandeis, G. (2013). Thermal evolution of an early magma ocean in interaction with the atmosphere. *Journal of Geophysical Research (Planets)*, 118, 1155–1176.

Matas, J., Bass, J., Ricard, Y., Mattern, E., Bukowinski, M.S.T. (2007). On the bulk composition of the lower mantle: Predictions and limitations from generalized inversion of radial seismic profiles. *Geophysical Journal International*, 170(2), 764–780.

Maurice, M., Tosi, N., Samuel, H., Plesa, A.-C., Hüttig, C., Breuer, D. (2017). Onset of solid-state mantle convection and mixing during magma ocean solidification. *Journal of Geophysical Research (Planets)*, 122(3), 577–598.

McDonough, W.F., Šrámek, O., Wipperfurth, S.A. (2020). Radiogenic power and geoneutrino luminosity of the Earth and other terrestrial bodies through time. *Geochemistry, Geophysics, Geosystems*, 21(7), e08865.

Meyer, B.S. and Clayton, D.D. (2000). Short-lived radioactivities and the birth of the sun. *Space Science Reviews*, 92, 133–152.

Monteux, J. and Arkani-Hamed, J. (2016). Scaling laws of impact induced shock pressure and particle velocity in planetary mantle. *Icarus*, 264, 246–256.

Monteux, J., Ricard, Y., Coltice, N., Dubuffet, F., Ulvrova, M. (2009). A model of metal-silicate separation on growing planets. *Earth and Planet. Sci. Lett.*, 287, 353–362.

Monteux, J., Andrault, D., Samuel, H. (2016). On the cooling of a deep terrestrial magma ocean. *Earth and Planetary Science Letters*, 448, 140–149.

Monteux, J., Andrault, D., Guitreau, M., Samuel, H., Demouchy, S. (2020). A mushy Earth's mantle for more than 500 Myr after the magma ocean solidification. *Geophysical Journal International*, 221(2), 1165–1181.

Nakajima, M. and Stevenson, D.J. (2015). Melting and mixing states of the Earth's mantle after the Moon-forming impact. *Earth and Planetary Science Letters*, 427, 286–295.

Nomura, R., Ozawa, H., Tateno, S., Hirose, K., Hernlund, J.W., Muto, S., Ishii, H., Hiraoka, N. (2011). Spin crossover and iron-rich silicate melt in the Earth's deep mantle. *Nature*, 472, 199–202.

Nomura, R., Hirose, K., Uesugi, K., Ohishi, Y., Tsuchiyama, A., Miyake, A., Ueno, Y. (2014). Low core-mantle boundary temperature inferred from the solidus of pyrolite. *Science*, 343(6170), 522–525.

Pierazzo, E., Vickery, A.M., Melosh, H.J. (1997). A reevaluation of impact melt production. *Icarus*, 127, 408–423.

Pierru, R., Pison, L., Mathieu, A., Gardés, E., Garbarino, G., Mezouar, M., Hennet, L., Andrault, D. (2022). Solidus melting of pyrolite and bridgmanite: Implication for the thermochemical state of the Earth's interior. *Earth and Planetary Science Letters*, 595, 117770.

Qaddah, B., Monteux, J., Le Bars, M. (2020). Thermal evolution of a metal drop falling in a less dense, more viscous fluid. *Physical Review Fluids*, 5(5), 053801.

Ray, R.D., Eanes, R.J., Lemoine, F.G. (2001). Constraints on energy dissipation in the Earth's body tide from satellite tracking and altimetry. *Geophysical Journal International*, 144(2), 471–480.

Reali, R., Van Orman, J.A., Pigott, J.S., Jackson, J.M., Boioli, F., Carrez, P., Cordier, P. (2019). The role of diffusion-driven pure climb creep on the rheology of bridgmanite under lower mantle conditions. *Scientific Reports*, 9, 2053.

Ricard, Y., Šrámek, O., Dubuffet, F. (2009). A multi-phase model of runaway core-mantle segregation in planetary embryos. *Earth and Planet. Sci. Lett.*, 284, 144–150.

Riley, G.N. and Kohlstedt, D.L. (1991). Kinetics of melt migration in upper mantle-type rocks. *Earth and Planetary Science Letters*, 105(4), 500–521.

Roche, P.-E. (2020). The ultimate state of convection: A unifying picture of very high Rayleigh numbers experiments. *New Journal of Physics*, 22(7), 073056.

Roscoe, R. (1952). The viscosity of suspensions of rigid spheres. *British Journal of Applied Physics*, 3(8), 267–269.

Rubie, D.C., Melosh, H.J., Reid, J.E., Liebske, C., Righter, K. (2003). Mechanisms of metal-silicate equilibration in the terrestrial magma ocean. *Earth and Planet. Sci. Lett.*, 205, 239–255.

Salvador, A., Massol, H., Davaille, A., Marcq, E., Sarda, P., Chassefière, E. (2017). The relative influence of H_2O and CO_2 on the primitive surface conditions and evolution of rocky planets. *Journal of Geophysical Research (Planets)*, 122(7), 1458–1486.

Samuel, H., Tackley, P.J., Evonuk, M. (2010). Heat partitioning in terrestrial planets during core formation by negative diapirism. *Earth and Planetary Science Letters*, 290, 13–19.

Schramm, D.N., Tera, F., Wasserburg, G.J. (1970). The isotopic abundance of ˇ26Mg and limits on ˇ26Al in the early Solar System. *Earth and Planetary Science Letters*, 10, 44–59.

Scott, T. and Kohlstedt, D.L. (2006). The effect of large melt fraction on the deformation behavior of peridotite. *Earth and Planetary Science Letters*, 246(3–4), 177–187.

Siggia, E.D. (1994). High rayleigh number convection. *Annual Review of Fluid Mechanics*, 26, 137–168.

Sleep, N.H., Zahnle, K.J., Lupu, R.E. (2014). Terrestrial aftermath of the Moon-forming impact. *Philosophical Transactions of the Royal Society of London Series A*, 372(2024), 20130172.

Solomatov, V.S. (2000). Fluid dynamics of a terrestrial magma ocean. In *Origin of the Earth and Moon*, Canup, R.M. and Righter, K. (eds). University of Arizona Press, Tuscson.

Solomatov, V.S. (2007). Magma oceans and primordial mantle differentiation. In *Treatise of Geophysics*, volume 9, Schubert, G. (ed.). Elsevier, Amsterdam.

Solomatov, V.S. and Stevenson, D.J. (1993). Suspension in convective layers and style of differentiation of a terrestrial magma ocean. *Journal of Geophysical Research*, 98, 5375–5390.

Stevenson, D.J. (2003). Planetary magnetic fields. *Earth and Planetary Science Letters*, 208, 1–2.

Tateno, S., Hirose, K., Ohishi, Y. (2014). Melting experiments on peridotite to lowermost mantle conditions. *Journal of Geophysical Research (Solid Earth)*, 119(6), 4684–4694.

Tonks, W.B. and Melosh, H.J. (1993). Magma ocean formation due to giant impacts. *J. Geophys. Res.*, 98, 5319–5333.

Touboul, M., Kleine, T., Bourdon, B., Palme, H., Wieler, R. (2007). Late formation and prolonged differentiation of the Moon inferred from W isotopes in lunar metals. *Nature*, 450, 1206–1209.

Ulvrová, M., Labrosse, S., Coltice, N., Råback, P., Tackley, P.J. (2012). Numerical modelling of convection interacting with a melting and solidification front: Application to the thermal evolution of the basal magma ocean. *Physics of the Earth and Planetary Interiors*, 206, 51–66.

Wiechert, U., Halliday, A.N., Lee, D.-C., Snyder, G.A., Taylor, L.A., Rumble, D. (2001). Oxygen isotopes and the moon-forming giant impact. *Science*, 294, 345–348.

Williams, G.E. (2000). Geological constraints on the Precambrian history of Earth's rotation and the Moon's orbit. *Reviews of Geophysics*, 38(1), 37–60.

Yoshino, T., Walter, M.J., Katsura, T. (2003). Core formation in planetesimals triggered by permeable flow. *Nature*, 422, 154–157.

Yu, S. and Garnero, E.J. (2018). Ultralow velocity zone locations: A global assessment. *Geochemistry, Geophysics, Geosystems*, 19(2), 396–414.

4

Hotspots, Large Igneous Provinces and Global Mantle Dynamics

Cinzia G. FARNETANI
IPGP, Université de Paris-Cité, France

4.1. Introduction

In the 1960s, the theory of plate tectonics provided a coherent framework to interpret volcanism at plate boundaries, namely, spreading ridges and subduction zones. At spreading ridges, the diverging plates enable mantle rocks to ascend at relatively shallow depth and melting occurs because of adiabatic decompression (McKenzie and Bickle 1988). Note that rocks melting beneath ridges do not need to be hotter than the surrounding mantle, and their estimated potential temperature[1] is ~1,370±50°C (Putirka 2005). At subduction zones, plate boundaries converge and the oceanic lithsophere bends and subducts. With increasing pressure, the fluids contained in the hydrated crustal rocks are released to the mantle wedge. These fluids play a

1 Pedagogic Box: Potential temperature T_p is the temperature that a volume of rock would have if adiabatically decompressed to the surface. Adiabatically means without heat exchange. The relation between real and potential temperature is given as: $T_r = T_p\, exp(g\, \alpha\, z/C_p)$, where α is the thermal expansion coefficient, z is depth and C_p is the specific heat at constant pressure.

Structure and Dynamics of the Earth's Interior 1,
coordinated by Julien MONTEUX.

key role for partial melting, because their presence decreases the rock's solidus temperature (Katz et al. 2003).

Plate tectonics, however, did not provide a mechanism to explain intraplate volcanism, like Hawaii, nor the excess of volcanism observed along parts of a spreading ridge, like in Iceland. In 1971, Jason Morgan proposed that hotspots are the surface expression of mantle plumes upwelling from the Earth's lowermost mantle (Morgan 1971). Mantle plumes can be defined as localized upwelling currents of solid rocks that are intrinsically hotter than the surrounding mantle. The estimated potential temperature for the Hawaiian plume is 1,620±55°C, whereas for the Iceland plume, the temperature is 1,535±80°C (Putirka 2005), showing that plumes have an excess temperature of order 150–250°C with respect to the surrounding mantle. A recent study by Bao et al. (2022) proposes lower excess temperatures for several plumes, but the most vigorous ones still have high excess temperatures, more specifically: $\Delta T_{Iceland}$=221°C, ΔT_{Hawaii}=171°C, ΔT_{Samoa}=189°C, $\Delta T_{Galapagos}$=185°C. The excess temperature is important because (a) it enables plumes to melt even below a thick lithosphere, (b) it provides the plume's buoyancy force necessary to ascend from the deep mantle, (c) it poses a first-order constraint on the source zone of mantle plumes (Stacy and Loper 1983). Indeed, the most plausible mechanism to generate volumes of hotter mantle is through a dynamic instability of a hot thermal boundary layer[2]. At present, geophysical observations support the existence of a hot thermal boundary layer at the base of the Earth's mantle. Across the seismically observed D" layer, 100–200 km thick, Jeanloz and Morris (1986) estimated a temperature increment of 800–1,000°C, based on the mismatch between adiabatic temperatures in the lower mantle and in the core.

Mantle plumes thus represent an exceptional and unique window into the deep mantle: by deciphering their "geochemical fingerprint", we can constrain the chemical evolution of the Earth's mantle over billions of years

2 Pedagogic box: Across a thermal boundary layer (TBL) heat transfer occurs by conduction, rather than by convection. Since conduction is an inefficient mechanism to transfer heat, the thermal gradient ($\Delta T/\Delta z$) across the TBL is high. The lithosphere is a cold TBL, with a $\Delta T \sim$ 1,300°C over a depth range Δz ~100–150 km. A hot TBL at the base of the mantle might have a ΔT of 800–1,000°C. The existence of a TBL at 660 km depth is unlikely, since the endothermic phase transition does not constitute a complete barrier to convection, so that heat can be advected through this depth range.

and the possible preservation of Hadean[3] mantle domains, whereas, by deciphering their "present-day seismological fingerprint", we can constrain geodynamic processes governing the interplay between plate tectonics, mantle convection and deep mantle structures (see a recent review by Koppers et al. (2021) and references therein). Moreover, the massive flood volcanism that punctuated the geologic history has been associated with the arrival of a mantle plume at sub-lithospheric depths. Plume magmatism causes the emplacement of large igneous provinces (LIPs) (Figure 4.1) and volcanic passive margins, such as those of the North Atlantic Volcanic Province or Etendeka-Paraná, which formed just before continental rifting. Plumes may trigger continental break-up (Storey 1995; Courtillot et al. 1999), and the volcanic out-gassing (Self et al. 2006; Ganino and Arndt 2009) caused palaeoenvironmental changes affecting life (Renne et al. 1995; Svensen et al. 2009; Hull et al. 2020).

4.2. Active hotspots today

How many hotspots are there on the Earth's surface today? According to Ito and van Keken (2007), the number ranges between 45 and 70. Hotspots younger than 100 Ma are generally active, although their vigor may vary considerably: older hotspots are either waning ($100<\text{age}<140$ Ma) or inactive ($\text{age}>150$ Ma). Here, we discuss four key features generally used to characterize a hotspot.

First, the onset of hotspot magmatism is often, but not always, marked by a Large Igneous Province, a term including continental flood basalts (CFBs) and oceanic plateaus (OPs) formed, respectively, in continental and oceanic setting. Estimated extruded volumes of CFBs are $1\text{–}2\times10^6$ km^3, with perhaps similar (but unknown) volumes of intrusive magmatism. OPs such as Ontong-Java are even more voluminous (Coffin and Eldholm 1994) but their origin might be related to a mantle plume interacting with a ridge triple junction. For both CFBs and OPs, the emplacement of large volumes of mafic (Mg and Fe rich) lavas lasted only a few millions of years, and "geologically speaking" LIP's magmatism is a rapid and transient event.

3 Pedagogic box: In the geologic timescale, the Hadean Eon starts with Earth's accretion 4.567 Ga ago and ends 4.0 Ga ago. The Eons are as follows: the Archean (4.0–2.5 Ga ago), the Proterozoic (2.5–0.541 Ga ago) and the Phanerozoic (from 541 Ma ago to present).

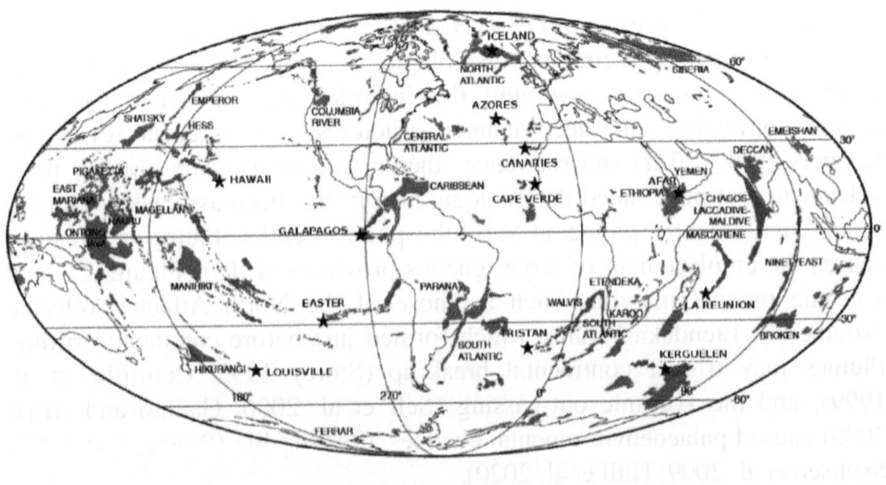

Figure 4.1. *Distribution of active hotspots (stars), LIPs (red) and the volcanic chain (blue). In the Atlantic ocean, the main hotspots are as follows: Iceland (0–62 Ma), the associated LIP is the North Atlantic Tertiary Igneous Province; Tristan da Cunha (0–125 Ma), the associated LIP is the Paraná-Etendeka Province. The Azores (0–20 Ma), Canaries (0–68 Ma) and Cape Verdes (0–Miocene) hotspots are not clearly associated with an LIP. In the Indian ocean, the main hotspots are as follows: Afar (0–30 Ma), the associated LIP is the Ethiopian Traps; La Réunion (0–65 Ma), the associated LIP is the Deccan Traps; Kerguelen (0.1–120 Ma), the associated LIP is the Kerguelen plateau. In the Pacific ocean, the main hotspots are as follows: Hawaii (0–76 Ma), the Emperor volcanic chain ends in the Aleutians and if an LIP existed, it has been subducted (Wei et al. 2020); Galápagos (0–90 Ma, possibly 140 Ma), the associated LIP is the Caribbean Plateau; Easter (0–26 Ma), maybe Tuamotu is the associated LIP; Louisville (1–77 Ma, possibly 120 Ma), maybe the Ontong-Java Plateau is the associated LIP. For some continental flood basalts, such as Emeishan (265 Ma), the Siberian Traps (251 Ma), and the Karoo-Ferrar Province (184 Ma), the associated hotspots are not clearly identified. For the Columbia River Flood basalt (17 Ma), the most likely hotspot is Yellowstone. Figure modified from Coffin and Eldholm (1994). For a color version of this figure, see www.iste.co.uk/monteux/earth1.zip*

Second, hotspots are often, but not always, characterized by an age progressive volcanic chain whose linear trend is consistent with the direction of plate motion. According to Ito and van Keken (2007), only 13 long-lived (>50 Ma) and eight short-lived (<20 Ma) volcanic chains are age-progressive, whereas all other volcanic chains lack a clear age progression. The Hawaiian–Emperor chain, an alignment of roughly 100 volcanoes, extends almost 6,000 km with a remarkable age progression. Other hotspots, like Cape Verde, lack a clear age progression because the hotspot position is very close to the Euler pole[4] describing the motion of the African plate (McNutt 1988). In simple words, this means that the plate velocity above the hotspot is close to zero. Finally, note that the two features presented above are consistent with the mantle plume initiation model by Richards et al. (1989), according to which the transient and episodic LIP magmatism corresponds to melting of a large plume head, whereas the subsequent hotspot activity is associated with the long-lasting and narrow plume tail.

Third, the active hotspot and the youngest part of the volcanic chain are characterized by a topographic swell. This is a region with an anomalously high topography (swell heights of 500–1,200 m) over lateral widths of 1,000–1,500 km, in the direction perpendicular to the volcanic chain (Crough 1983; Wessel 1993). The swell geometry is used to evaluate the plume buoyancy flux, a measure of the flux of buoyant material carried by the plume. According to Sleep (1990), the total buoyancy flux for all hotspots sums up to 55 Mg s^{-1}, whereas King and Adam (2014) found 22 Mg s^{-1}. Recently, Hoggard et al. (2020) proposed a new method, independent of plate velocity, and found 46 Mg s^{-1}, a value in global agreement with previous estimates.

Fourth, oceanic island basalts (OIBs), thought to be derived from plumes, are geochemically distinct and more diverse than mid-ocean ridge basalts (MORBs). As summarized by Hofmann (1997), OIBs have a different chemical composition in both trace and major elements (Schilling 1973) and a different isotopic composition (Hart et al. 1973). Hofmann and White (1982) first suggested that OIBs contain ancient recycled oceanic crust. OIBs also have a higher $^{3}He/^{4}He$ ratio[5] than MORB (Kurz et al. 1982; Farley and

4 Pedagogic box: The motion of a rigid body on a sphere is described as a rotation around a point called the Euler pole. The rotation axis goes through the center of the sphere and crosses the Earth's surface at the Euler pole. The Euler theorem relates the linear velocity of a point $v = \omega \times r$ to the rotation vector ω. At the Euler pole, $r = 0$, thus $v = 0$; see also Cox and Hart (1986).

5 Pedagogic box: The helium isotope ^{3}He is mostly primordial, whereas ^{4}He is primarily radiogenic, being produced by decay of ^{238}U, ^{235}U, ^{232}Th. High values of $^{3}He/^{4}He$, observed in some hotspots such as Hawaii, Iceland, Samoa and Galápagos, indicate high $^{3}He/(U+Th)$. The common interpretation is that the source region is ^{3}He rich, thus suggesting the involvement of a relatively undegassed, most primitive mantle. However, other interpretations exist: Coltice and Ricard (1999) suggested that a high $^{3}He/(U+Th)$ is due to a low (U+Th) in the source region. Finally, note that the $^{3}He/^{4}He$ ratio, called R, is often normalized over the atmospheric ratio Ra = 1.4 or 10^{-6}. The R/Ra ratio for MORB is ~8, whereas in some OIB can be up to 30.

Neroda 1998), which indicates the involvement of a relatively undegassed mantle component. Mundl-Petermeier et al. (2020) recently pointed out that plumes with high ^{3}He/^{4}He ratios also have a ^{182}W/^{184}W anomaly, this is particularly interesting because the ^{182}Hf $\rightarrow$ ^{182}W isotope system has a half-life of only 8.9 Myr (see Table 4.1).

Radioactive parent	Radiogenic daughter	Half-life (Myr)	Isotope ratio
^{147}Sm	^{143}Nd	106,000	^{143}Nd/^{144}Nd
^{87}Rb	^{87}Sr	48,800	^{87}Sr/^{86}Sr
^{238}U	^{206}Pb	4,469	^{206}Pb/^{204}Pb
^{182}Hf	^{182}W	8.9	^{182}W/^{184}W
^{129}I	^{129}Xe	15.7	^{129}Xe/^{136}Xe

Table 4.1. *Long-lived and short-lived isotopes*

4.3. Geochemistry of hotspot lavas: long-lived and short-lived isotope systems, what do they tell us?

During partial melting, elements in a solid rock behave differently, some go preferentially in the liquid phase, others stay in the solid. For an element i, the partition coefficient is $D^i = C^i_{sol}/C^i_{liq}$, where C^i_{sol} is the element concentrations in the solid and C^i_{liq} is the concentration in the liquid. Elements with $D^i > 1$ are called compatible, they concentrate in the solid phase, whereas incompatible elements ($D^i < 1$) concentrate in the liquid. Partial melting generates liquids that are enriched in incompatible elements, and this explains why the Earth's crust, which results from successive melting events, is enriched in incompatible elements (e.g. Rb, Th, U and K). The different behavior of elements upon melting provides a framework to understand the isotopic differences between MORBs and OIBs. In Figure 4.2, taken from Hofmann (2003), the apexes of the tetrahedron are associated with the isotope ratios ^{143}Nd/^{144}Nd, ^{87}Sr/^{86}Sr and ^{206}Pb/^{204}Pb (Table 4.1).

For the long-lived isotope system ^{147}Sm $\rightarrow$ ^{143}Nd (Figures 4.3(a) and (b)), the radioactive parent element ^{147}Sm behaves as a compatible element during partial melting. The residual solid, rich in ^{147}Sm, evolves over time (Gyr) into a depleted mantle, characterized by a high ^{143}Nd/^{144}Nd ratio. Instead, the partial melts, poor in radioactive parent element, evolve into crustal rocks, characterized by a low ^{143}Nd/^{144}Nd ratio. For the long-lived isotope system ^{87}Rb $\rightarrow$ ^{87}Sr (Figures 4.3(c) and (d)), the radioactive parent element Rb is more incompatible than Sr. The residual solid, poor in ^{87}Rb, evolves into a depleted mantle characterized by low ^{87}Sr/^{86}Sr ratio, whereas partial melts, rich in ^{87}Rb, evolve to crustal rocks with high ^{87}Sr/^{86}Sr.

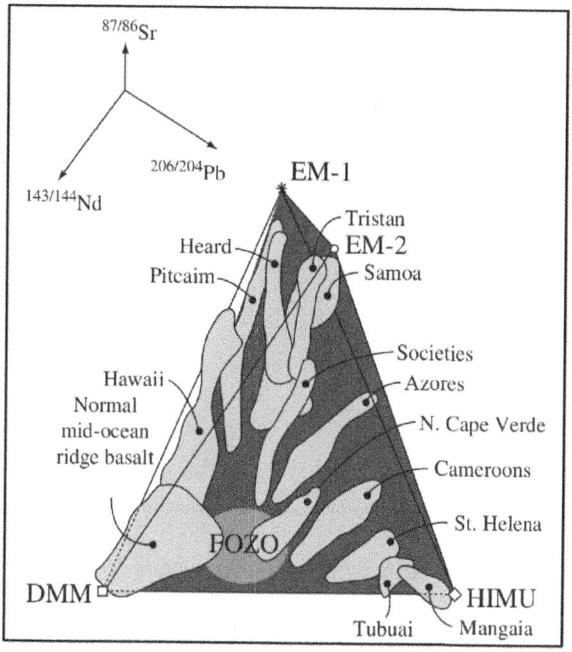

Figure 4.2. *The inset at the upper-left gives the direction in which ^{143}Nd/^{144}Nd, ^{87}Sr/^{86}Sr and ^{206}Pb/^{204}Pb increase. The apexes of the tetrahedron are as follows: depleted MORB mantle DMM, enriched mantle EM-1 and EM-2, and HIMU (high-μ, where μ is the ^{238}U/^{204}Pb ratio). In yellow, arrays of a large number of oceanic island basalts and the zone for normal mid-ocean ridge basalts. The pink area, labeled FOZO, corresponds to the so-called FOcalZone, thought to represent the composition of the deep mantle. Figure adapted from Hofmann (2003), after Hart et al. (1992). For a color version of this figure, see www.iste.co.uk/monteux/earth1.zip*

In the tetrahedron of Figure 4.2, the high values of each isotope ratio characterize an end-member composition, which is thought to reflect distinct mantle reservoirs: DMM for depleted MORB mantle, EM-1 and EM-2 for enriched mantle and HIMU which has a very radiogenic Pb isotopic composition, requiring a mantle source with high U/Pb for long time periods (1.5–2 Ga) because of the long half-life of the parent isotopes. Only a few hotspots have a strong HIMU component, namely, St Helena, and some hotspots of the Austral–Cook chain in the South Pacific. HIMU component

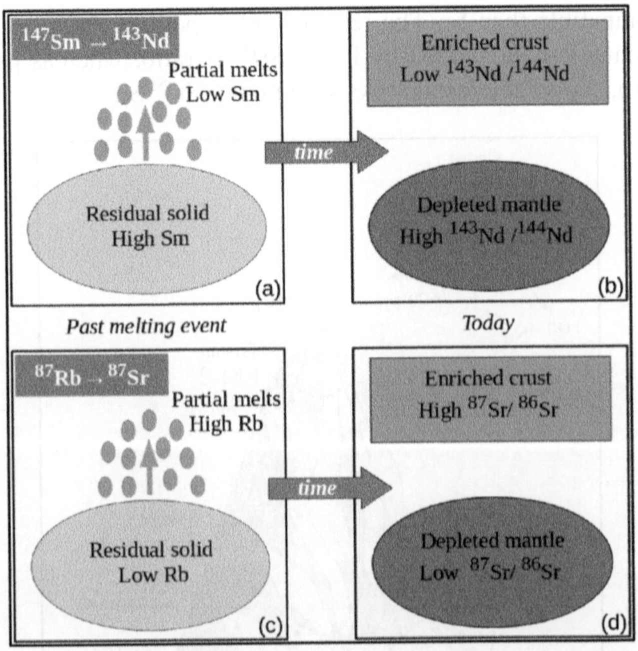

Figure 4.3. *Pedagogic scheme for the long-lived system $^{147}Sm \rightarrow {}^{143}Nd$ (a and b) and $^{87}Rb \rightarrow {}^{87}Sr$ (c and d). The aim of the figure is to understand why the depleted mantle (dark green) has high $^{143}Nd/^{144}Nd$ and low $^{87}Sr/^{86}Sr$, while the enriched crust (dark orange) has low $^{143}Nd/^{144}Nd$ and high $^{87}Sr/^{86}Sr$. Even if not shown in the figure, we note an interesting point: Partial melting generates a magma that has the same isotopic ratios as the source rock. For example, melts from the depleted mantle will have high $^{143}Nd/^{144}Nd$ and low $^{87}Sr/^{86}Sr$, simply because partial melting does not modify the isotopic ratio. Other processes, like magma mixing or assimilation of crustal rocks, can instead change the ratios. For a color version of this figure, see www.iste.co.uk/monteux/earth1.zip*

might reflect the presence of recycled oceanic crust, enriched in continental ^{238}U carried by sediments, or a recycled crust that lost part of its Pb during subduction. The EM2 component is evident in many hotspots, for example, Society, Marquesas and Samoa. This component indicates the involvement of continental material, later subducted in the form of sediments (the estimated amount of sediments in EM2 component is around 2%). The EM1 component may instead originate from recycling of delaminated subcontinental lithosphere and/or from recycling of subducted pelagic (i.e. open sea) sediments. The arrays of a large number of OIB indicate that most hotspot lavas are generated from isotopically "enriched" mantle sources, and this observation lead Hofmann and White (1982) to propose that mantle plumes carry ancient subducted oceanic crust. Subsequent geodynamic models confirmed that dense eclogitic crust does segregate in the lowermost mantle, before being entrained by plumes (Christensen and Hofmann 1994; Brandenburg et al. 2008; Nakagawa and Tackley 2014).

Recent advances in high-precision geochemical techniques enable to investigate short-lived isotopes, for example, ^{182}Hf $\rightarrow$ ^{182}W and ^{129}I $\rightarrow$ ^{129}Xe. Their half-life is so short (Table 4.1) that the radioactive parents, ^{182}Hf and ^{129}I, became extinct soon after the formation of the Solar System, $\sim$50 Myr and $\sim$80 Myr, respectively. Figure 4.4(a) shows that during core formation, W is moderately siderophile (i.e. goes into the metallic core), while Hf is lithophile, Figure 4.4(b) shows that in the presence of a magma ocean, Hf is more compatible than W (Deng and Stixrude 2021), so that the liquid magma ocean is slightly enriched in W. In Figure 4.4(c), we use μ^{182}W, defined as the deviation (in ppm) of ^{182}W/^{184}W of a sample from that of terrestrial standards. Since the terrestrial standard is similar to mantle rocks, the mantle has a μ^{182}W$\sim$0. Interestingly, material that formed an early magma ocean is expected to have slightly negative μ^{182}W. The Earth's core is a reservoir with very negative μ^{182}W$\sim$-220. At present, the observed μ^{182}W in MORBs and most OIBs is basically zero, but hotspots such as Hawaii, Samoa, Iceland and Galápagos have negative μ^{182}W, associated with high ^{3}He/^{4}He (Mundl et al. 2017; Mundl-Petermeier et al. 2020; Jackson et al. 2020). How to explain OIB's negative μ^{182}W anomalies? Schematically speaking there are two groups of models:

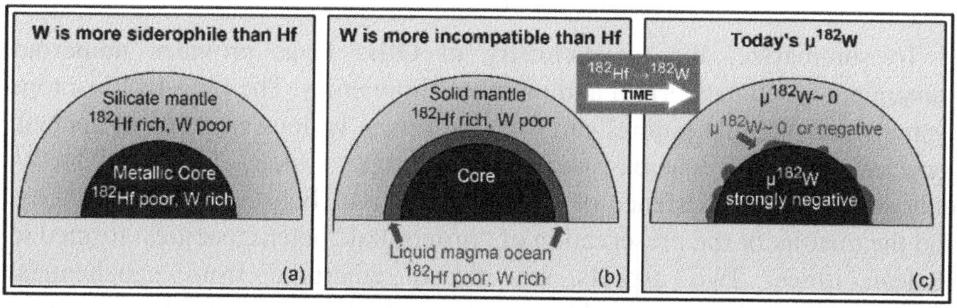

Figure 4.4. *Pedagogic scheme for the short-lived system ^{182}Hf $\rightarrow$ ^{182}W. While ^{182}Hf was extant: (a) ^{182}Hf stays preferentially in the mantle, rather than in the core. (b) ^{182}Hf stays preferentially in the solid phase, rather than in the liquid magma ocean. (c) At later times, the three reservoirs develop a distinct μ^{182}W. For a color version of this figure, see www.iste.co.uk/monteux/earth1.zip*

The first group invokes processes that occurred while ^{182}Hf was extant, such as the crystallization of a magma ocean (see Chapter 3). Quite independently of how crystallization proceeds (i.e. bottom up or not, see Labrosse et al. (2007); Maurice et al. (2017)), the residual liquid in the magma ocean becomes progressively enriched in incompatible elements, with a high iron content and a sub-chondritic Hf/W, leading to slightly negative μ^{182}W. A key geodynamic issue is how to preserve this "Enriched Magma-Ocean Residue" from vigorous convective mixing over 4 billion years. We still do not have an answer, but a compositional excess density seems to be required to segregate and store material in the deep mantle. In the literature, the term "mantle overturn" describes a large-scale mantle instability that brings dense material formed at relatively shallow depths to the bottom of the mantle, thereby leading to a stable compositional stratification (Brown et al. 2014).

The second group invokes processes that occurred after ^{182}Hf extinction, such as reactions between the mantle and the core (Rizo et al. 2019). These reactions may form a "Core-Mantle Equilibrated Reservoir" (Figure 4.5) chemically and isotopically equilibrated with the core (Mundl-Petermeier et al. 2020), and thus with a core-like, strongly negative μ^{182}W. The entrainment of only 0.3% of this reservoir could explain the most negative μ^{182}W ($\sim$-23) observed in OIBs (Mundl-Petermeier et al. 2020). The idea that diffusive transfer of W from the core to the lowermost mantle can create isotopic heterogeneities was recently supported by high-pressure experiments (Yoshino et al. 2020) showing that grain boundary diffusion of W is a fast and strongly temperature-dependent process.

To summarize, the geochemistry of OIBs lavas provides numerous constraints on plumes origin and on mantle dynamics. The long-lived isotope systems indicate that plumes carry to the surface various amounts of ancient, recycled material, whereas short-lived isotope systems, such as Hf-W, indicate either the existence of chemical reactions between the Earth's core and the mantle, or the preservation of "primordial" heterogeneities, formed in Hadean times. Two aspects are clearly emerging from geochemical observations of plume-derived hotspot lavas: (i) the case for a deep origin of mantle plumes is unequivocally strengthened and (ii) different isotopic systems need to be considered simultaneously and interpreted coherently. For example, Jackson et al. (2020) nicely showed that a simultaneous signature of high ^{3}He/^{4}He and negative μ^{182}W is best preserved in plumes lacking a strong signatures of recycled material (i.e. plumes with high ^{143}Nd/^{144}Nd and low ^{206}Pb/^{204}Pb).

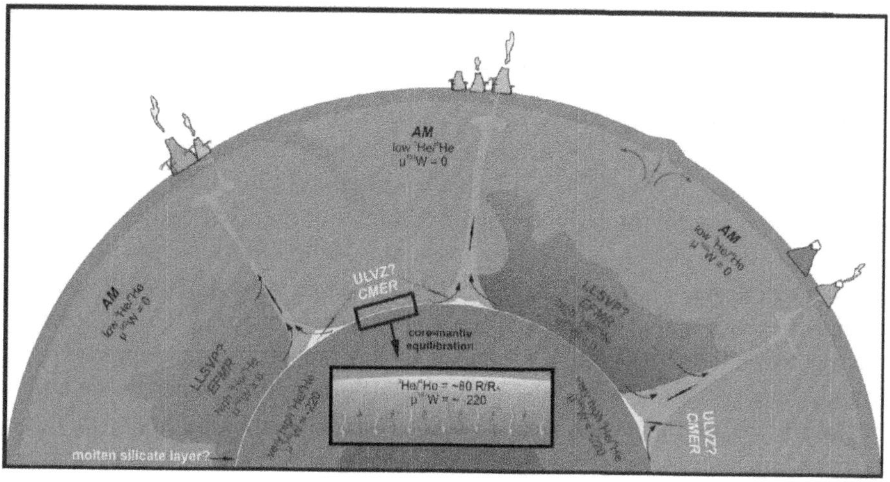

Figure 4.5. *From Mundl-Petermeier et al. (2020). CMER is the core-mantle equilibrated reservoir, and with strongly negative $\mu^{182}W$, it might correspond to the seismically observed ULVZs (yellow). EFMR is the early formed mantle reservoir, and with high $^3He/^4He$, it might correspond to the seismically observed LLSVPs (red). AM is the ambient mantle (violet). For a color version of this figure, see www.iste.co.uk/monteux/earth1.zip*

4.4. Seismic imaging below hotspots: To which extent do LLSVPs and ULVZs "feed" mantle plumes?

Ever since the plume concept has been proposed, the question addressed to seismologist was: Is there a deep seismic velocity anomaly[6] beneath hotspots? Montelli et al. (2006) first showed the existence of plume conduits below several hotspots using an innovative technique of finite-frequency tomography. Two main observations were as follows: (a) zones with a negative shear-wave velocity anomaly were broader than the expected narrow conduits, and (b) the depth extent of the seismic anomaly was quite variable, and beneath some

6 Pedagogic box: The energy released by an earthquake generates seismic waves that travel inside the Earth. After an earthquake, thousands of seismic stations record the arrival times of seismic waves. The observed arrival time is then compared with the predicted travel time, calculated using a reference model, such as PREM (Dziewonski and Anderson 1981). Any discrepancy between observed and predicted travel time is interpreted as a discrepancy between the "real" seismic velocity at depth, and the reference seismic velocity provided by PREM. Simply stated, slow (fast) seismic velocity anomalies are associated with hotter (colder) portions of the mantle.

hotspots the anomaly extended to the lowermost mantle; other hotspot instead lacked a clear continuity over all mantle depths.

Seismic tomography showed the existence of two low-velocity structures in the lower mantle below Africa and the central Pacific. The "Large Low Shear Velocity Provinces" (LLSVP), observed in both shear- and compressional-wave tomography, cover 25–30% of the CMB surface and rise 500–1,000 km above it (Ritsema et al. 1999; Masters et al. 2000; Romanowicz and Gung 2002). LLSVPs are often associated with mantle plumes beneath active hotspots (French and Romanowicz 2015); moreover, the reconstructed LIP position (obtained by calculating the plate motion back in time to the location where the LIP erupted) often corresponds to the edges of the LLSVPs (Burke and Torsvik 2004). LLSVP are likely hotter than the surrounding mantle, but there is still a debate about their nature, either purely thermal or thermo-chemical. Masters et al. (2000) observations of an anti-correlation between S-wave velocity and bulk sound speed[7] anomalies strengthened the likelihood of a distinct composition. From a geodynamic stand point, it is quite difficult to conceive that the Earth's mantle is purely thermal. Integrated over 4 Gyr, the mass of subducted oceanic crust represents 10–15% the mass of the mantle, and the subducted basaltic crust remains compositionally denser even in the lowermost mantle (Stixrude and Lithgow-Bertelloni 2012). Geodynamic models confirmed that dense recycled crust tends to pile-up at the base of the mantle, (Christensen and Hofmann 1994; Brandenburg et al. 2008) forming LLSVP-like structures. However, Deschamps et al. (2012) pointed out that LLSVPs may not consist of pure basaltic crust since in the lowermost mantle V_S increases with increasing fraction of MORB. The authors instead propose a more "primitive" composition enriched in iron by 3.0% and in (Mg,Fe)-perovskite by 20% compared to regular mantle. The last 10 years have seen the bloom of thermo-chemical geodynamic models, combining recycled crust and primitive material (Li et al. 2014; Nakagawa and Tackley 2014; Jones et al. 2021). The results confirm the heterogeneous nature of the LLSVPs and that plumes may be fed by LLSVPs material. However, even the latest geodynamic models do not capture the complex shape observed by the most recent tomographic studies. For example, for the African LLSVP,

7 Pedagogic box: The compressional wave velocity $V_P = \sqrt{(K + G3/4)/\rho}$ and the shear wave velocity $V_S = \sqrt{G/\rho}$ are combined together to calculate the bulk sound speed $\phi = \sqrt{V_P^2 - V_S^2 4/3}$, so that $\phi = \sqrt{K/\rho}$, where K is the modulus of incompressibility, G is the modulus of rigidity and ρ is the rock density.

Tsekhmistrenko et al. (2021) show that mantle upwellings are arranged in a tree-like structure: from a central, compact trunk below ~1,500 km depth, three branches tilt outwards and up toward various Indo-Austral hotspots. Similarly, for the Pacific LLSVP, Davaille and Romanowicz (2020) show the presence of well-separated, low-velocity conduits that extend vertically throughout most of the lower mantle.

Last, but not least, seismologists discovered the ultra low velocity zones (ULVZ), patches on the core-mantle boundary that exhibit reductions of S-wave velocity up to 30% and P-wave velocity up to 10% (Thorne and Garnero 2004). ULVZs are 10–40 km high and have a lateral extent of hundreds of kilometers, but up to 1,000 km for the Mega-ULVZs (Cottaar and Romanowicz 2012). Their density increase can be up to 10% (Rost et al. 2005). Several models have been proposed to explain this high value, and they include core-mantle reaction, dense silicate melts, iron-enriched oxides (see the review by McNamara (2019) and references therein), and even relics of banded iron formations subducted 2–3 Ga ago (Dobson and Brodholt 2005).

What has been presented is certainly incomplete, but the aim is to underline the following: the seismically observed LLSVPs and ULVZs are increasingly considered by geochemists as repositories of enriched mantle components either related to the early differentiation of the Earth, or to core-mantle reactions, or to the accumulation of subducted material. For example, Mundl-Petermeier et al. (2020) explain hotspots lavas with negative μ^{182}W and high ^{3}He/^{4}He, by invoking a “Core-Mantle Equilibrated Reservoir”, thought to correspond to the ULVZs, and an “Early Formed Mantle Reservoir”, with high ^{3}He/^{4}He, thought to correspond to the LLSVPs (Figure 4.5). These new ideas open geodynamic questions on how, and how much, compositionally denser material can be brought to the surface by plumes.

4.5. Plumes in the convecting mantle

The Earth’s mantle is a “thermal system” cooled from above, heated from within through the decay of radioactive elements, and heated from below through core heat flux (see Chapter 5). The Earth’s mantle is bounded by two thermal boundary layers, both dynamically unstable: the cold lithosphere sinks and subducts, while instabilities close to the core-mantle boundary give rise to hot rising plumes. The existence of a hot thermal boundary layer capable of generating plumes requires that the overlying mantle is efficiently cooled. This a plausible assumption for the last 1–2 Gyr, but one can ask about earlier times, when the mantle was 200–300°C hotter than today and when mantle convection was in a stagnant-lid regime, a configuration that inhibited and delayed the onset of plume activity. We do not have a definitive answer, although an early plume activity is supported by eruption of komatiites through most of the Archean (Arndt et al. 2008).

Our view of mantle plumes has evolved from the "classical" mushroom-shaped thermal plume to the more complex thermo-chemical plumes, and let us examine both of them: For purely thermal plume (Figure 4.6(a)), the only source of buoyancy is represented by their excess temperature with respect to the surrounding mantle. If rock viscosity is strongly temperature dependent, then the hot plume develops a large head and a narrow tail (Griffiths 1990; Olson et al. 1993). Across the plume tail, the vertical velocity decreases exponentially with radial distance, a velocity profile that causes high shear of the upwelling material (Farnetani and Hofmann 2009). If rock viscosity is weakly temperature dependent, or if other factors like the grain size increase the viscosity, then the plume morphology is "spout-like", with a similar radius for head and tail. Obviously, during upwelling the plume shape can be modified, for example, at levels associated with a global viscosity change (e.g. at 660 km depth) or because "mantle wind" tilts the plume.

Thermo-chemical plumes have two sources of buoyancy: thermal and compositional. The thermal and compositional density differences are often in competition, for example, eclogitic recycled crust is denser than the pyrolitic mantle. Dannberg and Sobolev (2015) estimated that a plume with an excess temperature of 250°C can entrain ~20% of dense eclogite before becoming neutrally buoyant. Concerning the composition of primitive mantle, or of portions of the mantle that interacted with the metallic core, there is still much debate, however deep-seated, iron rich mantle is expected to be compositionally denser. As shown by fluid dynamic laboratory experiments (Davaille 1999; Kumagai et al. 2008) and by numerical simulations (Tackley 1998; Farnetani and Samuel 2005), thermo-chemical plumes are broad and have an irregular dome-like shape. The internal dynamics is complex since governed by both positive-thermal and negative-compositional buoyancy forces: some parts of the plume conduit might sink while other parts slowly upwell, or the whole plume conduit might "collapse" and sink because of insufficient buoyancy (see also the review by Koppers et al. (2021) and references therein). Furthermore, compositional heterogeneities might have a different rheology, causing even more complex dynamics in the plume conduit (Farnetani et al. 2018).

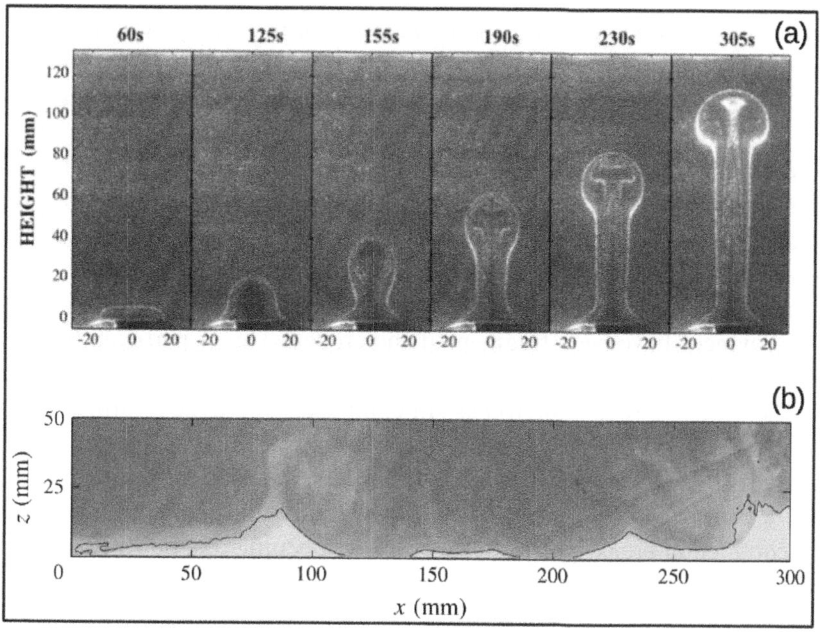

Figure 4.6. *(a) Image of a laboratory experiment featuring a purely thermal plume generated by a heated patch at the base of the tank (Davaille et al. 2011). (b) Image of a laboratory experiment in a microwave oven from Limare et al. (2019). Over time, a deep seated layer of internally heated material (yellow) becomes unstable and forms cusps. Thermo-chemical plumes rise from these cusps. Note that in the experiment there is no basal heating. For a color version of this figure, see www.iste.co.uk/monteux/earth1.zip*

Last but not least, laboratory experiments by Limare et al. (2015) and Limare et al. (2019) showed that plumes can be generated even in the absence of a bottom-heated thermal boundary layer. A deep-seated layer, compositionally enriched in heat producing elements (U, Th, K) can develop an excess temperature sufficient to form plumes (Figure 4.6(b)). The dynamic instability leading to plume formation results from the interaction between the internally heated layer and cold, subducted material.

4.6. Large igneous provinces

This section gives a general overview of the main LIPs, while it is out of my purpose to deal with the vast literature specific of each LIP (see, e.g. Coffin and Eldholm (1994); Eldholm and Coffin (2000); Wignall (2001); Courtillot and Renne (2003); Ito and van Keken (2007); Ernst and Youbi (2017) and references therein). A CFB province comprises hundreds of individual lava flows, with volumes of hundreds and up to thousands of cubic kilometers. These lavas are primarily tholeiitic basalts, but a small volume fraction (5–20%) can be siliceous (rhyolitic-trachytic) and sometimes carbonatites and kimberlites are associated with CFBs (Ernst and Bell 2010; Ernst and Youbi 2017). The total extruded volume of a CFB can be of order 1 Mkm^3. The surface extent ($\sim$1 Mkm^2) might be underestimated since most CFB provinces have been fragmented and eroded. This is particularly true for PreCambrian CFB, where the entire volcanic sequence has been eroded and only the intrusive rocks outcrop as sill complexes and dyke swarms. CFB are unique geological features because the eruption of large volumes of magma occurred on a short time scale of order 1 Myr. However, the magmatic activity can last longer when CFBs are spatially and temporally associated with continental break-up. In such a case, the opening of a new ocean basin is marked by the formation of volcanic margins and sea-ward dipping layers (Hopper et al. 2003).

4.6.1. *Phanerozoic continental flood basalts*

The *Columbia River Basalts* is the youngest CFB, with an eruption age of 16.6–15.3 Ma and is also the smallest province, with an estimated erupted volume of 0.17 Mkm^3. A remarkable feature is that individual lava flows can be up to 600 km long and their estimated volumes is 2,000 km^3. According to Lange (2002), a high volatiles concentration (H_2O and CO_2 up to 4%) can explain the large individual flow volumes and the ability of the dense Fe-rich lava to erupt through 20 km of continental crust.

The *Yemen–Ethiopia traps*, formed $\sim$30 Ma ago, presently outcrop in Ethiopia, Eritrea, Djibouti and Yemen, over a surface of 0.7 Mkm^2. Mafic magmas ($\sim$0.35 Mkm^3, but probably up to 1 Mkm^3) were followed by rhyolitic pyroclastic rocks (Courtillot and Renne 2003; Ito and van Keken 2007). The flood volcanism lasted 1–2 Myr and it preceded the onset of extension and oceanization along the East African Rift ($\sim$23 Ma) and in the Gulf of Aden ($\sim$26 Ma).

In the *North Atlantic Tertiary Volcanic Province* volcanism started ~61 Ma ago, with voluminous (~2 Mkm^3) tholeiitic eruptions in West and Southeast Greenland, British Isles and Baffin island, over a surface if 1.3 Mkm^2 (Abbott and Isley 2002). This first pulse of volcanic activity lasted 1–2 Ma and it is interpreted as the arrival of a mantle plume head beneath central Greenland. The second pulse (~56 Ma ago) was even more voluminous (> 2 Mkm^3) and it is linked to continental rifting between East Greenland and Europe (56–52 Ma). The rifted margins of the North Atlantic are characterized by a thick crust (> 15 km) with seaward-dipping seismic reflection sequences. Oceanic drilling, both off Norway and Greenland, revealed that the sequences are made of basalts erupted subaerially, or at shallow water depths, indicative of regional uplift (see Hopper et al. (2003) and references therein).

The *Deccan Traps* (India) were emplaced ~66 Ma ago, within 1 Myr around the Cretaceous–Paleogene boundary (e.g. Courtillot and Renne (2003); Sprain et al. (2019) and references therein). The end of the Cretaceous corresponds to a major global extinction, where 75% of all species lost, including non-avian dinosaurs. Although the temporal link between some CFB provinces and major extinctions appears to be robust (e.g. the Siberian Traps at 251 Ma corresponds to the Permian-Triassic extinction; the Central Atlantic Magmatic Province at 201 Ma corresponds to the Triassic–Jurassic extinction), there has been much debate concerning the Deccan Traps. This is because a collision between the Earth and an asteroid of more than 10 km in diameter occurred in the Yucatan peninsula (Mexico) precisely at the Cretaceous–Paleogene boundary (Swisher III. et al. 1992; Schulte et al. 2010). Therefore, the Deccan volcanism began prior to the impact and straddled it in time. The Deccan Traps (Figure 4.7(a)) were emplaced on thick Archean crust and their estimated volume is of order 2 Mkm^3 (Coffin and Eldholm 1994). Their current surface is of order 0.5 Mkm^2, but erosion and continental break-up between India and the Seychelles (~63 Ma) significantly reduced the surface. The geochemical fingerprint of lavas least contaminated by the continental lithosphere is similar to oceanic hotspot basalts and the major element contents suggests high temperature of melting, suggestive of a plume involvement.

The *Paraná traps* (South America) and the smaller Etendeka traps (Namibia and Angola) are conjugate volcanic fields split by the break-up of South America and Africa. The Paraná province covers 1.2 Mkm^2 and an even larger extent is supported by extensive dyke swarms. The Etendeka province covers 0.08 Mkm^2 and is very similar to the Paraná basalts in terms of eruptive history, petrology and geochemistry (Renne et al. 1996; Courtillot and Renne 2003). Minor eruptions in the northwest of the Paraná basin started at 135–138 Ma, whereas most of the trap volcanism occurred at 133±1 Ma (Renne et al. 1992) and lasted ~0.6 Ma. Trap emplacement was followed by continental break-up, as evidenced by rift-related dykes in coastal Brazil, beginning around 128 Ma, and by a persistent (128–120 Ma) magmatism along the coast, eventually leading to the opening of the South Atlantic Ocean, and the formation of the Rio Grande and Walvis OPs.

Figure 4.7. *(a) Photo of the Deccan Traps by Mark Richards. (b) Siberian flood-basalt flows in Putorana, Taymyr Peninsula (Wignall 2011). For a color version of this figure, see www.iste.co.uk/monteux/earth1.zip*

The *Karoo province* in South Africa and the *Farrar basalts*, a linear belt extending for 4,000 km along the Transantarctic mountains (Antarctica), are the remnants of a large volcanic province, with an estimated volume of 2.5 Mkm^3. The bulk of the province formed 183±1 Ma ago, but the igneous activity continued for a few millions of years (e.g. the Okavango dyke swarm in Botswana has an age of ~179 Ma), suggesting a long-lived volcanic activity, with an apparent south-to-north migration (Jourdan et al. 2005). The province was then split apart by the break-up of Gondwana and the subsequent seafloor spreading (~156 Ma).

Remnants of the *Central Atlantic Magmatic Province (CAMP)* presently outcrop in North Africa, North America and Central South America. The province might have covered a surface of 7 Mkm^2, with a volume exceeding 2 Mkm^3 (Marzoli et al. 1999) but what is visible today are only giant dyke swarms, sills and some lava flows. The magmatic activity started around 201 Ma, had a peak at 200 Ma, and went on to 197 Ma, being associated with the continental break-up of Pangea and the opening of the central Atlantic Ocean. According to Abbott and Isley (2002), a triple junction formed during CAMP extrusion, leading to the divergence of three continental blocks: Africa/Europe, North America and South America. The lack of a clear plume track led some authors to question the involvement of a mantle plume, for example, Holbrook and Kelemen (1993) suggest that the offshore portion of the CAMP, along the rifted continental margins, formed without a plume. As discussed later, there might be a temporal link between the earliest CAMP volcanism and the mass extinction at the Triassic–Jurassic boundary (80% of species lost).

The *Siberian Traps*, with an estimated volume of 3 Mkm^3, erupted 251 Ma ago (Kamo et al. 1996) a time period corresponding to the Permo-Triassic mass extinction (95% of species lost). At present, the traps are exposed over an area of only 0.4 Mkm^2, but their subsurface extent is much bigger, as indicated by drill holes (Reichow et al. 2002). Several geological evidences indicate that the original surface of the traps might have been in excess of 4 Mkm^2. The eruption of the enormous Siberian Traps probably lasted only 1 Myr, and this rapid sequence is attested by the lack of sediments or paleosols between flows (see Courtillot and Renne (2003) and references therein). Individual lava flows (~150 m thick) extend for hundreds of kilometers, and pile up to give an average thickness of ~1,000 m (Figure 4.7(b)). Lavas of the Siberian traps were rich in volcanic gases, namely, CO_2 and SO_2 (Kamo et al. 2003), as supported by many kimberlite pipes, east of the traps, which correspond to an early, highly explosive phase of the volcanism (Kravchinsky et al. 2002). Moreover, the basal parts of the Siberian traps contain mafic tuffs, attesting an elevated gas content. A high volatile content might be due to interactions between the ascending magma and the sedimentary sequence, particularly rich in anhydrite and coal. As discussed in the next section, the impact of the Siberian traps on the environment might be linked to explosive eruptions and the release of sulfur and carbon in the atmosphere. Finally, the strong link between LIPs and Ni-Cu-PGE (platinum group element) ore deposits is evident in the Noril'sk area (e.g. Ernst et al. (2005) and references therein).

The *Emeishan province* outcrops in southwestern China, over an area of only 0.3 Mkm^2. This surface is much smaller than the original area because the Emeishan traps have been severely fragmented, deformed and eroded since their formation, ca. 258 Ma ago. The observed thickness of the traps is of order 1–2 km, leading to an estimated volume of $\sim$1 Mkm^3. The eruption was preceded by a kilometer-scale uplift (He et al. 2003), suggesting dynamic uplift induced by an ascending mantle plume (e.g. Xu et al. (2004) and references therein). Zhou et al. (2002) point out the temporal link between the formation of the Emeishan traps and end-Guadalupian mass extinction, while Ganino and Arndt (2009) investigate the environment impact, as discussed in the next section.

Finally, we note that other LIPs have been recently discovered, for example, the 275 Ma Bachu LIP in Central China (Zhang et al. (2010) and references therein) and the 300 Ma Skagerrak-Centered LIP in NW Europe (Torsvik et al. 2008).

4.6.2. *Phanerozoic oceanic plateaus*

Here, some of the most important OPs are presented, selecting those that are more clearly linked to a hotspot. Since the oldest oceanic lithosphere on Earth is only 180 Ma old, the time record for OPs is much shorter than for CFB provinces. Old OPs have been subducted or they may be accreted as allochtonous terranes, for example, the Wrangellia terrane (Figure 4.8), outcropping in British Columbia and in SE Alaska, is a fragment of a 230 Ma old OP (Richards et al. 1991; Greene et al. 2008).

The Cretaceous *Caribbean plateau* now forms the Caribbean plate, which is characterized by a thickened oceanic crust (up to 20 km). Moreover, fragments of the plateau were accreted along the Pacific margins of Central and South America (e.g. Guatemala, Costa-Rica, Colombia and Equador). According to Hoernle et al. (2004), the volcanic province covers an area of $\sim$2,500 km (E-W) and 1,300 km (N-S), with an estimated volume of order 4 Mkm^3 and a surface of 1.8 Mkm^2 (Abbott and Isley 2002). The often quoted age of the plateau is 89.5$\pm$0.3 Ma, however a full range of dates

(91–88 Ma) suggests a longer volcanic history (Sinton et al. 1998). The plateau is generally linked to the birth of the Galápagos hotspot, although Hoernle et al. (2004) question the classical "plume-head" model, based on the protracted volcanic activity.

Figure 4.8. *Wrangell Mountains, Alaska. The photo shows 1,000 m of continuous flood basalt stratigraphy. This is an accreted fragment of an ancient oceanic plateau. The yellow line marks the contact between Nikolai basalts and the overlying Chitistone Limestone (Greene et al. 2008). For a color version of this figure, see www.iste.co.uk/monteux/earth1.zip*

The *Kerguelen plateau*, in the southern Indian Ocean, has a long and complex tectono-magmatic history: the Southern part of the Plateau formed at 119 Ma and was shortly followed by the emplacement of the Rajmahal Traps in India at 117–118 Ma. The Central part of the Kerguelen Plateau formed by 110 Ma and Broken Ridge by 95 Ma (Coffin et al. 2002). This protracted volcanic activity created a province with an extent of more than 2 Mkm^2 and a volume of 15–24 Mkm^3 (Coffin and Eldholm 1994) that was split apart by the Southeast Indian ridge, only 40 My ago. The eruption rate of the Kerguelen hotspot decreased from 1.7 km^3/year during plateau formation to 0.18 km^3/year during the emplacement of the Ninetyeast Ridge (82–43 Ma). A "hotspot track" formed as the young oceanic lithosphere of the Indian plate moved over the hotspot. The magmatic activity of the hotspot persisted until 0.1 Ma ago (Nicolaysen et al. 2000) but with very low eruption rates (0.009 km^3/year).

The *Ontong-Java plateau*, in the southwestern Pacific ocean, is the largest LIP. Its surface extent reaches 2 Mkm^2 and its mean crustal thickness is estimated at 36 km, leading to a total volume of the order 44 Mkm^3 (see Eldholm and Coffin (2000) and references therein). It has been suggested that the Ontong-Java plateau is only a fragment of what was originally an even larger edifice that included Manihiki and Hikurangi Plateaus (Taylor 2006; Hoernle et al. 2010). Arndt and Davaille (2013) indicate that the emplacement of these enormous OPs might have increased the rate of seafloor spreading as well as the subduction rates around the Pacific rim. Although there is no clear consensus on the age and duration of the magmatism, the Ontong-Java plateau likely formed around 122 Ma ago, with a major pulse lasting 1–2 Myr, whereas the second pulse (at 90 Ma) was minor. The eruptions were submarine and occurred on very young oceanic lithosphere, possibly close to a ridge. This is supported by petrological models (Fitton and Godard 2004) showing that primary magmas formed by 30% melting of a peridotitic source with a hot (1,500°C) potential temperature. The basalts of the Ontong-Java plateau share many characteristics of typical MORBs, however we point out that the eight ocean drill holes "scratch the surface" of the plateau, since they penetrate only 300 m into the volcanic sequence. Luckily, accreted fragments enable to study the internal structure of the plateau. During the last 6 Ma, collision of the Ontong-Java plateau with the Solomon arc caused folding and uplift of the southern margin of the plateau. At present, fragments of the "unsubductable" Ontong-Java plateau form a $\sim$3 km thick section of basaltic rocks exposed in the Solomon Islands, more specifically in Malaita, Santa Isabel and San Cristobal (Petterson 2004; Kroenke et al. 2004).

4.6.3. *Older LIPs and radiating dyke swarms*

Tracing Precambrian LIPs is not an easy task! Oceanic LIPs have either been subducted or are preserved as obducted fragments but their LIP origin may not even be recognized. Concerning Precambrian continental LIPs, most often their basaltic component has been eroded and only the intrusive component is now exposed (i.e. dykes, sills and eroded magma chambers which form mafic and ultramafic layered intrusions). Another difficulty with

Precambrian LIPs is to define the full extent of the province, because this requires continental reconstructions of separated crustal blocks that were formerly adjacent. Key evidence for the identification of Precambrian LIPs are radiating dyke swarms that might extend laterally over distances of up to 2,000 km. Obviously, some swarms have been fragmented by plate tectonics, and at present about 30 giant radiating swarms have been identified (Abbott and Isley 2002), including the huge Mackenzie swarm in Canada (Ernst and Baragar 1992).

A full record of Precambrian LIPs is given by Ernst and Youbi (2017) and references therein, and here I will cite only some of the most important events: The *Central Iapetus Magmatic Province (CIMP)*, a multi-pulsed event that occurred between 615 and 555 Ma, is widespread in eastern Laurentia, Baltica and West Africa. The *Franklin LIP* event in northern Canada and NW Greenland, of age 723±3 Ma, is now recognized to be even larger, with its extension into southern Siberia as the Irkutsk LIP. This magmatic event might be linked to the Rodinia supercontinent break-up, which was punctuated by episodic plume events at ~825 Ma (the Gairdner-Amata dyke swarm in central and SE-Australia), at ~780 Ma (the radiating Gunbarrel dyke swarms in western Laurentia) and 750–720 Ma (e.g. Li et al. (2008) and references therein). The *Mackenzie LIP*, with a single short duration pulse at 1,269±2 Ma is the largest LIP event of the Canadian Shield. Ernst and Baragar (1992) studied the huge, fan-shaped, Mackenzie mafic dyke swarm (Figure 4.9), which extends for hundreds of kilometers away from the region of plume-centered flood volcanism (i.e. the Coppermine River basalt). The average dyke width is greater than 10 m, and the length attests that magma traveled horizontally at least 2,100 km. Ernst et al. (2008) propose that the Mackenzie event, together with older ones (1,380 and 1,460 Ma) can be associated with the break-up of the late-Paleoproterozoic supercontinent, Nuna (Columbia).

Older LIPs do exist, for example, the *North-American Circum-Superior LIP* (1,864–1,885 Ma), a belt of mafic-ultramafic rocks and carbonatite complexes surrounding much of the Superior craton (Ciborowski et al. 2017). Archean LIPs include the Fortescue flood basalt (Pilbara, Australia, 2,770 Ma) and the Ventersdorp flood basalt (Kaapvaal, South Africa, 2,713 Ma). Moreover, the association of tholeites and komatiites (indicative of high temperatures) in metamorphic greenstone belts may attest Archean LIP events, as the 2,730–2,700 Ma magmatic event, recognized in many cratons around the world (see Ernst et al. (2005) and references therein). The details of this subject can be found by starting from Ernst and Youbi (2017),

Ernst et al. (2008), Ernst et al. (2005), Abbott and Isley (2002) and references therein. What is important here is to point out that LIP events occurred throughout the geological times and likely contributed to the formation of the continental crust.

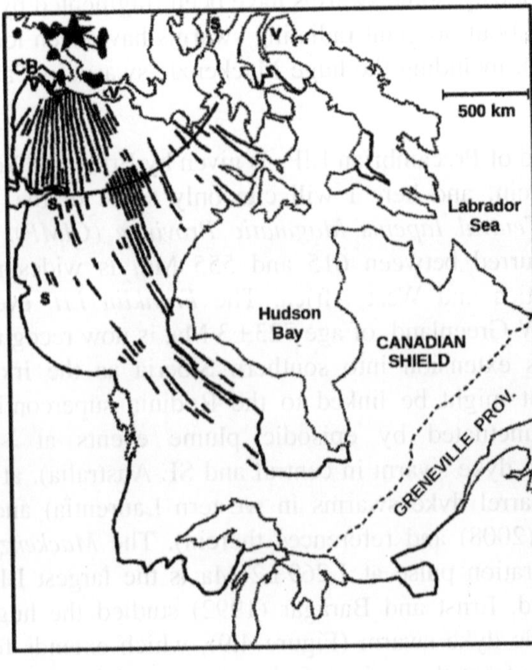

Figure 4.9. *Figure from Ernst et al. (2001) showing the giant Mackenzie dyke swarm. Layered intrusions, mostly interpreted on the basis of geophysics, are in black. Coppermine River basalts (CB) at the north end of the Mackenzie swarm might indicate the plume center. The arc denotes the transition from proximal, radiating dykes, to the distal, sub-linear part of the swarm. S for sills and V for flood basalts*

4.7. Environmental effects of Phanerozoic large igneous provinces

Catastrophic changes in fauna and flora occurred in the Phanerozoic Era, and five events are considered as major extinction, the so-called "Big Five": end-Cretaceous (66 Ma; 75% of all species lost), end-Triassic (201 Ma; 80% of species lost), end-Permian (251 Ma; 95% of species lost), end-Devonian (~375 Ma; 75% of species lost) and end-Ordovician (444 Ma; 85% of species lost). Here, I synthesize the effects of LIPs magmatism on the environment, in particular how atmospheric stress contributed to the collapse of biota diversity. The discussion of other processes causing environmental changes (e.g. meteorite impacts, changes in solar luminosity and plate-boundaries tectonics) is outside the focus of this chapter. CFBs inject heat, volcanic fluids and gases into the atmosphere, leading to a deterioration of the global environmental conditions for the living species (Figure 4.10).

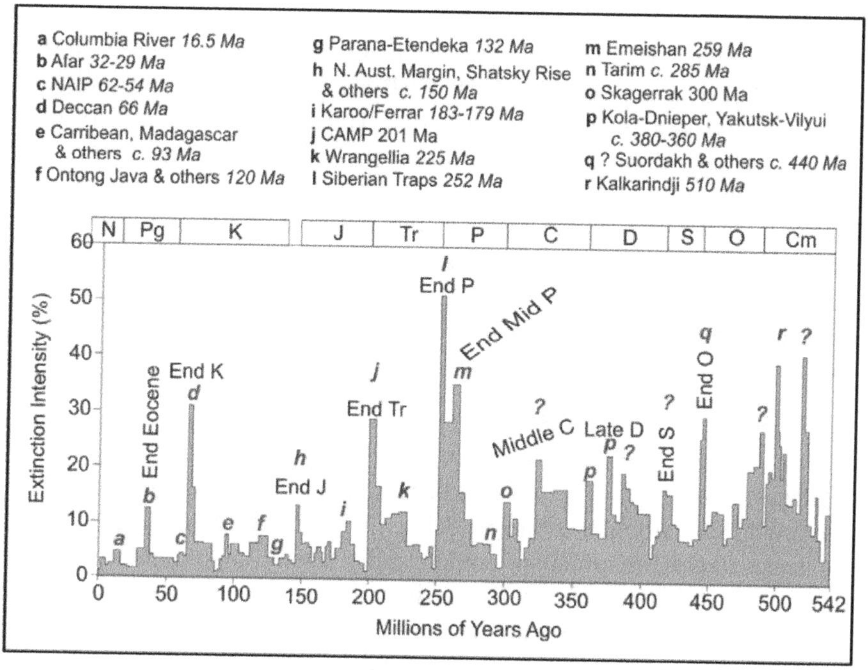

Figure 4.10. *The Extinction intensity represents the fraction of marine genera that are present in each time interval, but do not exist in the next one. High values mark important extinctions. The red letters correspond to LIP events, listed above the figure. Figure adapted from (Wignall 2011). For a color version of this figure, see www.iste.co.uk/monteux/earth1.zip*

Petrological studies of CFB provinces suggest that 3.5–6.5$\times 10^9$ kg SO_2 are released every 1 km^3 of basaltic magma (Mather and Schmidt 2021). Moreover, sulfur concentrations vary from province to province (Callegaro et al. 2014), from 1,900 ppm in Deccan and CAMP, both associated with major extinction events, to 800 ppm in Paraná-Etendeka, a CFB without major environmental effects. Estimates of released CO_2 are challenging, because of the low solubility of CO_2 in magmas, but are in the order of 10^{16}–10^{17} kg of CO_2 for an entire province. The effect of volcanic CO_2 injection in the atmosphere can also be revealed by shifts (i.e. a lowering) in the carbon isotopic composition, since mantle-derived CO_2 has a negative (-4 $^0/_{00}$) $\delta^{13}C$[8].

4.7.1. *Factors triggering global warming and global cooling*

The prolonged release of volcanogenic greenhouse gases can induce global warming. At the end of Permian, the equatorial sea surface temperature (SST) increased by 5–10°C (Black et al. (2018) and references therein), likely as a result of the Siberian Traps emplacement. Hull et al. (2020) showed that a large pulse of CO_2 released before the Cretaceous–Paleogene boundary, caused a 2°C warming, likely as a result of the early volcanic activity in Deccan. For the CAMP province, we lack a precise value of the temperature variation, but excursion of the light carbon isotope ($\delta^{13}C_{org}$) attests of increased atmospheric partial pressure in CO_2 (pCO_2) (Hesselbo et al. 2002), synchronous to the onset of CAMP volcanism at the Triassic–Jurassic boundary.

Greenhouse gases released in the atmosphere during LIP emplacement may have various sources: (a) an extrusive source, associated with volcanic CO_2 degassed during tuffaceous and basaltic eruption. Ernst and Bell (2010) pointed out that carbonatites are associated with many LIPs, suggesting that carbonatite-alkaline magmatism could rapidly release high amounts of CO_2 and SO_2 into the atmosphere; (b) an intrusive source, associated with the heating of country rocks by the intruded magmas (Ganino and Arndt 2009). Reactions between thick magmatic sills and volatile-rich sedimentary strata may favor gases accumulation, in particular if the strata have a low permeability. Volatiles, such as CO_2, sulfur and halocarbons, are then released through hydrothermal vent complexes, which attest of an explosive activity. Vent complexes have been discovered in association with the Siberian Traps, the North Atlantic Igneous Province, Karoo and the Central Atlantic Magmatic Province (Svensen et al. 2007, 2009). The case of the Siberian Traps is particularly interesting, since the sills intruded organic rich shale and petroleum bearing evaporites, which are sources of halocarbons, CH_4 and CO_2. Metamorphism of organic matter and petroleum by intrusions could have generated $>10^5$ Gt of CO_2 (Svensen et al. 2009). (c) Assimilation

8 Pedagogic box: The difference between $^{13}C/^{12}C$ of the sample and $^{13}C/^{12}C$ of a standard reference is normalized over $^{13}C/^{12}C$ of the reference. This ratio, multiplied by 1,000, gives the $\delta^{13}C$ (in per mil). Living organisms prefer the light isotope ^{12}C and have a negative (-25$^0/_{00}$) $\delta^{13}C$.

of a metasomatized continental lithosphere may also be a source of volatiles (Broadley et al. 2018), in particular if ancient subduction enriched the subcontinental lithospheric mantle. Note that degassing of volatiles originated from the interaction between intrusive magmas and country rocks, also called "Cryptic degassing" (McKay et al. 2014) can occur at different times with respect to volcanic degassing accompanying basalt eruptions. Finally, the increase in atmospheric greenhouse gases, and the associated temperature increase, may drive positive feedback mechanisms possibly leading to a runaway greenhouse effect (Wignall 2001). For example, the destabilization of marine clathrates (i.e. methane hydrate sediments with very negative -$60^0/_{00}$ δ^{13}C), or the sudden release of methane from permafrost, or increased wildfires, all cause the release of even more greenhouse gases. Last but not least, habitat loss and see-level changes, due to global warming, intensify the ecological stress.

A key process triggering global cooling is the draw down of atmospheric CO_2 by intense chemical weathering of silicate rocks. Chemical weathering occurs during major orogenic episodes of continental collision (Goddéris et al. 2014) and can be particularly efficient on basaltic rocks in wet, hot climates. Therefore, the consumption of atmospheric CO_2 by silicate weathering is an important mechanism to prevent the runaway greenhouse conditions described above. Injection of sulfur dioxide in the atmosphere is another process triggering global cooling, because sulfur forms sunlight blocking sulfate aerosols. Black et al. (2018) modeled the competing effects of sulfur and carbon emissions associated with the emplacement of the Siberian Traps. Their results show that multiple episodes of intense magmatism caused repeated swings between a cooler "sulfur climate mode" and warmer "carbon climate mode".

4.7.2. *Oceanic anoxia events: ocean acidification*

Because of the interchange between the atmosphere and the oceans, the formation of a CFB can affect the marine record in several ways: (a) global warming increases the likelihood for the marine water column to become stratified and for the deep waters to become anoxic. The term oceanic anoxia events (OAEs) refers to periods of oceanic environmental crisis characterized by low, or oxygen-absent, deep-ocean conditions. Black shales are indicative of global OAEs and they can be linked to some LIPs (e.g. see Pearce et al. (2008); Ernst and Youbi (2017) and references therein). (b) Ocean acidification, caused by acid rains, rich in carbonic and sulfuric acids, has devastating consequences on calcified marine biota. The fossil record shows major loss of calcifying organisms during the events of ocean acidification. These events can be linked to some LIPs, for example, the Siberian Traps, the CAMP and the Deccan Traps (Ernst and Youbi 2017). Sobolev et al. (2011) point out that Pre-Mid-Mesozoic oceans were more vulnerable to crisis due to acidification because they lacked deep-sea, unlithified, carbonate sediments that act as a buffer against acidification.

4.7.3. *Ozone depletion from halogen emissions: release of toxic metals*

Melt inclusion studies on the Deccan Traps and the Siberian Traps yield a range of total halogen (HCl and HF) emissions of order 10^{14}–10^{16} kg (Mather and Schmidt 2021). More specifically, degassing of the Siberian Traps released 6.3–7.8$\times 10^{15}$ kg of S, 3.4–8.7 $\times 10^{15}$ kg of Cl and 7.1–13.6$\times 10^{15}$ kg of F (Black et al. 2012). Halogens injected in the atmosphere disrupt the ozone, and indeed an ozone depletion was an important factor for the end-Permian mass extinction (Broadley et al. 2018). Metal toxicity, in particular of mercury and arsenic, contributes to environmental crises. High mercury levels have been recognized in association with a least five LIP events, including the CAMP and Karoo–Ferrar (Percival et al. 2015), the Siberian and the Deccan Traps and the Ontong-Java plateau (see Ernst and Youbi (2017) and references therein).

For OPs, the environmental effect might be less important because magmas cross the relatively thin oceanic lithosphere, poor in volatiles. Moreover, the enormous mass of ocean water is expected to act as a buffer (Ernst and Youbi 2017), although injection of volcanic CO_2 would still lead to ocean anoxia and acidification. Indeed, Kerr (2005) showed the temporal coincidence of global oceanic anoxia, black shale deposition and the formation of OPs. The most extensive concentrations of organic-rich black shales in the geological record occur in the Aptian (124–112 Ma), a time interval corresponding to the formation of the Ontong Java Plateau, Hess Rise, Manihiki Plateau, the East Mariana and Nauru basins, plus a significant proportion of the Kerguelen Plateau. The Aptian shales, besides indicating oceanic anoxia, also show a sharp decrease in δ^{13}C, suggesting the dissociation of methane hydrates, either triggered by global warming or by tectonic events related to plume uplift. Finally, the emplacement of OPs can cause sea-level fluctuations and perturb oceanic currents, possibly promoting stratification of the water column and anoxia.

4.8. Concluding remarks

This chapter focused on present-day hotspots and on Phanerozoic LIPs. For active hotspots, their geochemical fingerprint and what tomography reveals about the seismic velocity anomaly at depth was briefly presented. For LIPs, the most important Phanerozoic CFB and OPs were touched upon, paying attention to the effect of LIP volcanism on the atmosphere, the oceans and on life. Older LIPs do exist, but these provinces have been eroded, fragmented and split apart by plate tectonics. Nonetheless, the existence of Archean LIPs attests that episodes of massive magmatism punctuated the geologic times during the last 2.5 Ga.

LIPs and most of active hotspots are associated with mantle plumes. Therefore, by studying the geochemistry of intraplate, plume-related, volcanism, we can attempt to constrain the dynamics and the composition of the deep mantle. We have known for decades that plumes carry to the surface ancient (1–2 Ga old) recycled crust, but what has been revealed by short-lived isotopes is totally new: some hotspots, like Hawaii, Samoa, Iceland and Galápagos, have a negative μ^{182}W anomaly. This observation poses several unanswered questions: (a) To understand present-day mantle composition, should not we start from its initial condition, that is, from the crystallization of the magma ocean? (b) Is it possible to form mantle reservoirs that are "preserved" from vigorous mantle stirring over time-scales of 4 Ga? (c) Are the negative μ^{182}W anomalies caused by chemical interactions between the metallic core and the mantle? The above questions target the deep mantle and the way heterogeneities are formed and stored before being entrained by plumes. Other unanswered questions target the shallow mantle, in particular the thermo-mechanical interactions between the plume and the overlying lithosphere: for example, (d) How does partial melt ascend and interact with lithospheric and crustal rocks? (e) Why are some LIPs associated with continental break-up, whereas others are not? (f) After partial melting, the residual rocks are expected to be compositionally lighter (since Fe depleted) and more viscous (since fluids depleted). What is their dynamic role in the transition from massive LIPs magmatism to the reduced but long-lived hotspot magmatism?

In the future years, interdisciplinary studies constrained by mineral physics results, geodynamic modeling (see Chapter 1) and observations from geochemistry (see Chapter 6) and seismology will hopefully provide an answer.

4.9. References

Abbott, D.H. and Isley, A.E. (2002). The intensity, occurrence, and duration of superplume events and eras over geological time. *J. Geodynamics*, 34, 265–307.

Arndt, N. and Davaille, A. (2013). Episodic Earth evolution. *Tectonophysics*, 609, 661–674.

Arndt, N., Lesher, C.M., Barnes, S.J. (2008). *Komatiite*. Cambridge University Press, Cambridge.

Bao, X., Lithgow-Bertelloni, C.R., Jackson, M.G., Romanowicz, B. (2022). On the relative temperatures of Earth's volcanic hotspots and mid-ocean ridges. *Science*, 375, 57–61.

Black, B.A., Elkins-Tanton, L.T., Rowe, M.C., Peate, I.U. (2012). Magnitude and consequences of volatile release from the Siberian Traps. *Earth Planet. Sci. Lett.*, 317–318, 363–373.

Black, B.A., Neely, R.R., Lamarque, J.F., Elkins-Tanton, L.T., Kiehl, J.T., Shields, C.A., Mills, M.J., Bardeen, C. (2018). Systemic swings in end Permian climate from Siberian Traps carbon and sulfur outgassing. *Nature Geoscience*, 11, 949–954.

Brandenburg, J.P., Hauri, E.H., van Keken, P.E., Ballentine, C.J. (2008). A multiple-system study of the geochemical evolution of the mantle with force-balanced plates and thermochemical effects. *Earth Planet. Sci. Lett.*, 276, 1–13.

Broadley, M.W., Barry, P.H., Ballentine, C.J., Taylor, L.A., Burgess, R. (2018). End Permian extinction amplified by plume induced release of recycled lithospheric volatiles. *Nature Geoscience*, 11, 682–687.

Brown, S.M., Elkins-Tanton, L.T., Walker, R.J. (2014). Effects of magma ocean crystallization and overturn on the development of 142Nd and 182W isotopic heterogeneities in the primordial mantle. *Earth Planet. Sci. Lett.*, 408, 319–330.

Burke, K. and Torsvik, T.H. (2004). Derivation of large igneous provinces of the past 200 million years from long-term heterogeneities in the deep mantle. *Earth Planet. Sci. Lett.*, 227, 531–538.

Callegaro, S., Baker, D.R., De Min, A., Marzoli, A., Geraki, K., Bertrand, H., Viti, C., Nestola, F. (2014). Microanalyses link sulfur from large igneous provinces and Mesozoic mass extinctions. *Geology*, 42, 895–898.

Christensen, U.R. and Hofmann, A.W. (1994). Segregation of subducted oceanic crust in the convecting mantle. *J. Geophys. Res.*, 99, 19867–19884.

Ciborowski, T.J.R., Minifie, M.J., Kerr, A.C., Ernst, R.E., Baragar, B., Millar, I.L. (2017). A mantle plume origin for the palaeoproterozoic circum-superior large igneous province. *Precambrian Research*, 294, 189–213.

Coffin, M.F. and Eldholm, O. (1994). Large igneous provinces: Crustal structure, dimensions, and external consequences. *Rev. Geophys.*, 32, 1–36.

Coffin, M.F., Pringle, M.S., Duncan, R.A., Gladczenko, T.P., Storey, M., Muller, R.D., Gahagan, L.A. (2002). Kerguelen hotspot magma output since 130 Ma. *J. Petrology*, 43, 1121–1139.

Coltice, N. and Ricard, Y. (1999). Geochemical observations and one layer mantle convection. *Earth Planet. Sci. Lett.*, 174, 125–137.

Cottaar, S. and Romanowicz, B. (2012). An unsually large ULVZ at the base of the mantle near Hawaii. *Earth Planet. Sci. Lett.*, 355, 213–222.

Courtillot, V.E. and Renne, P.R. (2003). On the ages of flood basalt events. *Comptes Rendus Geoscience*, 335, 113–140.

Courtillot, V.E., Jaupart, C., Manighetti, I., Tapponnier, P., Besse, J. (1999). On causal links between flood basalts and continental breakup. *Earth Planet. Sci. Lett.*, 166, 177–195.

Cox, A. and Hart, R.B. (1986). *Plate Tectonics: How It Works*. Blackwell Scientific Publications, Oxford.

Crough, S.T. (1983). Hotspot swells. *Annu. Rev. Earth Planet. Sci.*, 11, 165–193.

Dannberg, J. and Sobolev, S.V. (2015). Low-buoyancy thermochemical plumes resolve controversy of classical mantle plume concept. *Nat. Commun.*, 6, 6960.

Davaille, A. (1999). Simultaneous generation of hotspots and superswells by convection in a heterogeneous planetary mantle. *Nature*, 402, 756–760.

Davaille, A. and Romanowicz, B. (2020). Deflating the LLSVPs: Bundles of mantle thermochemical plumes rather than thick stagnant "piles". *Tectonics*, 39. doi: 10.1029/2020TC006265.

Davaille, A., Limare, A., Touitou, F., Kumagai, I., Vatteville, J. (2011). Anatomy of a laminar starting thermal plume at high Prandtl number. *Exp. Fluids*, 50, 285–300.

Deng, J. and Stixrude, L. (2021). Deep fractionation of Hf in a solidifying magma ocean and its implications for tungsten isotopic heterogeneities in the mantle. *Earth Planet. Sci. Lett.*, 562, 116873.

Deschamps, F., Cobden, L., Tackley, P.J. (2012). The primitive nature of large low shear-wave velocity provinces. *Earth Planet. Sci. Lett.*, 349–350, 198–208.

Dobson, D.P. and Brodholt, J.P. (2005). Subducted banded iron formations as a source of ultralow-velocity zones at the core-mantle boundary. *Nature*, 434, 371–374.

Dziewonski, A.M. and Anderson, D.L. (1981). Preliminary reference Earth model. *Phys. Earth Planet. Inter.*, 25, 297–356.

Eldholm, O. and Coffin, M.F. (2000). Large igneous provinces and plate tectonics. In *The History and Dynamics of Global Plate Motions*, Richards, M.A., Gordon, R.G., van der Hilst, R.D. (eds). John Wiley & Sons, New York.

Ernst, R.E. and Baragar, W.R.A. (1992). Evidence from magnetic fabric for the flow pattern in the Mackenzie giant radiating dyke swarm. *Nature*, 356, 511–513.

Ernst, R.E. and Bell, K. (2010). Large igneous provinces (LIPs) and carbonatites. *Miner. Petrol.*, 98, 55–76.

Ernst, R.E. and Youbi, N. (2017). How large igneous provinces affect global climate, sometimes cause mass extinctions, and represent natural markers in the geological record. *Palaeogeography, Palaeoclimatology, Palaeoecology*, 478, 30–52.

Ernst, R.E., Grosfils, E.B., Mége, D. (2001). Giant dike swarms: Earth, Venus, and Mars. *Annu. Rev. Earth Planet. Sci.*, 29, 489–534.

Ernst, R.E., Buchan, K.L., Campbell, I.H. (2005). Frontiers in large igneous province research. *Lithos*, 79, 271–297.

Ernst, R.E., Wingate, M.T.D., Buchan, K.L., Li, Z.X. (2008). Global record of 1600-700 Ma large igneous provinces (LIPs): Implications for the reconstruction of the proposed Nuna (Columbia) and Rodinia supercontinents. *Precambrian Research*, 160, 159–178.

Farley, K.A. and Neroda, E. (1998). Noble gases in the Earth's mantle. *Annu. Rev. Earth Planet. Sci.*, 26, 189–218.

Farnetani, C.G. and Hofmann, A.W. (2009). Dynamics and internal structure of a lower mantle plume conduit. *Earth Planet. Sci. Lett.*, 282, 314–322.

Farnetani, C.G. and Samuel, H. (2005). Beyond the thermal plume paradigm. *Geophys. Res. Lett.*, 32, L07311.

Farnetani, C.G., Hofmann, A.W., Duvernay, T., Limare, A. (2018). Dynamics of rheological heterogeneities in mantle plumes. *Earth Planet. Sci. Lett.*, 499, 74–82.

Fitton, J.G. and Godard, M. (2004). Origin and evolution of magmas on the Ontong Java Plateau. In *Origin and evolution of the Ontong Java Plateau*, Fitton, J.G., Mahoney, J.J., Wallace, P.J., Saunders, A.D. (eds). Geological Society, London.

French, S.W. and Romanowicz, B. (2015). Broad plumes rooted at the base of the Earth's mantle beneath major hotspots. *Nature*, 525, 95–99.

Ganino, C. and Arndt, N.T. (2009). Climate changes caused by degassing of sediments during the emplacement of large igneous provinces. *Geology*, 37, 323–326.

Goddéris, Y., Donnadieu, Y., Le Hir, G., Lefebvre, V., Nardin, E. (2014). The role of palaeogeography in the Phanerozoic history of atmospheric CO_2 and climate. *Earth Sci. Rev.*, 128, 122–138.

Greene, A.R., Scoates, J.S., Weis, D. (2008). Wrangellia flood basalts in Alaska: A record of plume-lithosphere interaction in a Late Triassic accreted oceanic plateau. *Geochem. Geophys. Geosyst.*, 9(12), Q12004.

Griffiths, R.G. and Campbell, H. (1990). Stirring and structure in mantle starting plumes. *Earth Planet. Sci. Lett.*, 99, 66–78.

Hart, S.R., Schilling, J.-G., Powell, J.L. (1973). Basalts from Iceland and along the Reykjanes Ridge: Sr isotope geochemistry. *Nature*, 246, 104–107.

Hart, S.R., Hauri, E.H., Oschmann, L.A., Whitehead, J.A. (1992). Mantle plumes and entrainment: Isotopic evidence. *Science*, 256, 517–520.

He, B., Xu, Y.G., Chung, S.L., Xiao, L., Wang, Y. (2003). Sedimentary evidence for a rapid, kilometer-scale crustal doming prior to the eruption of the Emeishan flood basalts. *Earth. Planet. Sci. Lett.*, 213, 391–405.

Hesselbo, S.P., Robinson, S.A., Surlyk, F., Piasecki, S. (2002). Terrestrial and marine extinction at the Triassic-Jurassic boundary synchronized with major carbon-cycle perturbation: A link to initiation of massive volcanism? *Geology*, 30, 251–254.

Hoernle, K., Hauff, F., van den Bogaard, P. (2004). 70 m.y. history (139-69 Ma) for the Caribbean large igneous province. *Geology*, 32, 697–700.

Hoernle, K., Hauff, F., van den Bogaard, P., Werner, R., Mortimer, N., Geldmacher, J., Garbe-Schonberg, D., Davy, B. (2010). Age and geochemistry of volcanic rocks from the Hikurangi and Manihiki oceanic plateaus. *Geochimica et Cosmochimica Acta*, 74, 7196–7219.

Hofmann, A.W. (1997). Mantle geochemistry: The message from oceanic volcanism. *Nature*, 385, 219–229.

Hofmann, A.W. (2003). Sampling mantle heterogeneity through oceanic basalts: Isotopes and trace elements. In *Treatise On Geochemistry: The Mantle and Core*, Carlson, R.W., Holland, H.D., Turekian, K.K. (eds). Elsevier, Oxford.

Hofmann, A.W. and White, W.M. (1982). Mantle plumes from ancient oceanic-crust, *Earth Planet. Sci. Lett.*, 57, 421–436.

Hoggard, M.J. and Parnell-Turner, R., White, N. (2020). Hotspots and mantle plumes revisited: Towards reconciling the mantle heat transfer discrepancy. *Earth Planet. Sci. Lett.*, 542, 116317.

Holbrook, W.S. and Kelemen, P.B. (1993). Large igneous province on the United-States Atlantic margin and implications for magmatism during continental breakup. *Nature*, 364, 433–436.

Hopper, J.R., Dahl-Jensen, T., Holbrook, W.S., Larsen, H.C., Lizarralde, D., Korenaga, J., Kent, G.M., Kelemen, P.B. (2003). Structure of the SE Greenland margin from seismic reflection and refraction data: Implications for nascent spreading center subsidence and asymmetric crustal accretion during North Atlantic opening. *J. Geophys. Res.*, 108(B5), 2269.

Hull, P.M., Bornemann, A., Penman, D.E., Henehan, M.J., Norris, R.D., Wilson, P.A., Blum, P., Alegret, L. Batenburg, S.J., Bown, P.R. et al. (2020). On impact and volcanism across the Cretaceous-Paleogene boundary. *Science*, 367, 266–272.

Ito, G. and van Keken, P.E. (2007). Hotspots and melting anomalies. In *Treatise on Geophysics: Mantle Dynamics*, Bercovici, D. (ed.). Elsevier, Amsterdam.

Jackson, M.G., Blichert-Toft, J., Halldorsson, S.A., Mundl-Petermeier, A. et al. (2020). Ancient helium and tungsten isotopic signatures preserved in mantle domains least modified by crustal recycling. *Proc. Natl. Acad. Sci. USA*, 117, 30993–31001.

Jeanloz, R. and Morris, S. (1986). Temperature distribution in the crust and in the mantle. *Ann. Rev. Earth Planet. Sci.*, 14, 377–415.

Jones, T.D., Sime, N., van Keken, P.E. (2021). Burying Earth's primitive mantle in the slab graveyard. *Geochem. Geophys. Geosyst.* doi: 10.1029/2020GC009396.

Jourdan, F., Féraud, G., Bertrand, H., Kampunzu, A.B., Tshoso, G., Watkeys, M.K., LeGall, B. (2005). Karoo large igneous province: Brevity, origin, and relation to mass extinction questioned by new Ar-40/Ar-39 age data. *Geology*, 33, 745–748.

Kamo, S.L., Czamanske, G.K., Krogh, T.E. (1996). A minimum U-Pb age for Siberian flood-basalt volcanism. *Geochim. et Cosmochim. Acta*, 60, 3505–3511.

Kamo, S.L., Czamanske, G.K., Amelin, Y., Fedorenko, V.A., Davis, D.W., Trofimov, V.R. (2003). Rapid eruption of Siberian flood-volcanic rocks and evidence for coincidence with the Permian-Triassic boundary and mass extinction at 251 Ma. *Earth Planet. Sci. Lett.*, 214, 75–91.

Katz, R.F., Spiegelman, M., Langmuir, C.H. (2003). A new parameterization of hydrous mantle melting. *Geochem. Geophys. Geosyst.*, 4(9). 1073.

Kerr, A.C. (2005). Oceanic LIPs: The kiss of death. *Elements*, 1, 289–292.

King, S.D. and Adam, C. (2014). Hotspot swells revisited. *Phys. Earth Planet. Inter.*, 235, 66–83.

Koppers, A.A.P., Becker, T.W., Jackson, M.G., Konrad, K., Muller, R.D., Romanowicz, B., Steinberger, B., Whittaker, J.M. (2021). Mantle plumes and their role in Earth processes. *Nature Reviews Earth and Environment*, 2, 382–401.

Kravchinsky, V.A., Konstantinov, K.M., Courtillot, V., Savrasov, J.I., Valet, J.-P., Chernyi, S., Mishenin, S., Parasotka, B. (2002). Paleomagnetism of East Siberian traps and kimberlites: Two new poles and paleogeographic reconstructions at about 360 and 250 Ma. *Geophys. J. Int.*, 148, 1–33.

Kroenke, L.W., Wessel, P., Sterling, A. (2004). Motion of the Ontong Java Plateau in the hot-spot frame of reference: 122 Ma-present. In *Origin and Evolution of the Ontong Java Plateau*, Fitton, J.G., Mahoney, J.J., Wallace, P.J., Saunders, A.D. (eds). Geological Society, London.

Kumagai, I., Davaille, A., Kurita, K., Stutzmann, E. (2008). Mantle plumes: Thin, fat, successful, or failing? Constraints to explain hot spot volcanism through time and space. *Geophys. Res. Lett.*, 35, L16301.

Kurz, M.D., Jenkins, W.J., Hart, S.R. (1982). Helium isotopic systematics of oceanic islands and mantle heterogeneity. *Nature*, 297, 43–47.

Labrosse, S., Hernlund, J., Coltice, N. (2007). A crystallizing dense magma ocean at the base of the Earth's mantle. *Nature*, 450, 866–869.

Lange, R.A. (2002). Constraints on the preeruptive volatile concentrations in the Columbia River flood basalts. *Geology*, 30, 179–182.

Li, Z.X., Bogdanova, S.V., Collins, A.S., Davidson, A., De Waele, B., Ernst, R.E., Fitzsimons, I.C.W., Fuck, R.A., Gladkochub, D.P. et al. (2008). Assembly, configuration, and break-up history of Rodinia: A synthesis. *Precambrian Research*, 160, 179–210.

Li, M., McNamara, A.K., Garnero, E.J. (2014). Chemical complexity of hotspots caused by cycling oceanic crust through mantle reservoirs. *Nature Geoscience*, 7, 366–370.

Limare, A., Vilella, K., DiGiuseppe, E., Farnetani, C.G., Kaminski, E., Surducan, E., Surducan, V., Neamtu, C., Fourel, L., Jaupart, C. (2015). Microwave-heating laboratory experiments for planetary mantle convection. *J. Fluid Mech.*, 777, 50–67.

Limare, A., Jaupart, C., Kaminski, E., Fourel, L., Farnetani, C.G. (2019). Convection in an internally heated stratified heterogeneous reservoir. *J. Fluid Mech.*, 870, 67–105.

Marzoli, A., Renne, P.R., Piccirillo, E.M., Ernesto, M., Bellieni, G., De Min, A. (1999). Extensive 200-million-year-old continental flood basalts of the Central Atlantic magmatic province. *Science*, 284, 616–618.

Masters, G., Laske, G., Bolton, H., Dziewonski, A. (2000). The relative behavior of shear velocity, bulk sound speed, and compressional velocity in the mantle: Implications for chemical and thermal structure. In *Mineral Physics and Tomography from the Atomic to the Global Scale*, Karato, S.I. et al. (eds). American Geophysical Union, Washington, D.C.

Mather, T.A. and Schmidt, A. (2021). Environmental effects of volcanic volatile fluxes from subaerial large igneous provinces. In *Large Igneous Provinces: A Driver of Global Environmental and Biotic Changes*, Ernst, R.E., Alex Dickson, A., Bekker, A. (eds). The American Geophysical Union and John Wiley and Sons, Washington, D.C./Hoboken, NJ.

Maurice, M., Tosi, N., Samuel, H., Plesa, A.-C., Huttig, C., Breuer, D. (2017). Onset of solid-state mantle convection and mixing during magma ocean solidification. *J. Geophys. Res. Planets*, 122, 577–598.

McKay, D.I.A., Tyrrell, T., Wilson, P.A., Foster, G.L. (2014). Estimating the impact of the cryptic degassing of Large Igneous Provinces: A mid-Miocene case-study. *Earth Planet. Sci. Lett.*, 403, 254–262.

McKenzie, D. and Bickle, M.J. (1988). The volume and composition of melt generated by extension of the lithosphere. *J. Petrology*, 29, 625–679.

McNamara, A.K. (2019). A review of large low shear velocity provinces and ultra low velocity zones. *Tectonophysics*, 760, 199–220.

McNutt, M. (1988). Thermal and mechanical properties of the Cape Verde Rise. *J. Geophys. Res.*, 93, 2784–2794.

Montelli, R., Nolet, G., Dahlen, F.A., Masters, G. (2006). A catalogue of deep mantle plumes: New results from finite-frequency tomography. *Geochem. Geophys. Geosyst.*, 7, Q11007.

Morgan, W.J. (1971). Convection plumes in the lower mantle. *Nature*, 230, 42–43.

Mundl, A., Touboul, M., Jackson, M.G., Day, J.M.D., Kurz, M.D., Lekic, V., Helz, R.T., Walker, R.J. (2017). Tungsten-182 heterogeneity in modern ocean island basalts. *Science*, 356, 66–69.

Mundl-Petermeier, A., Walker, R.J., Fischer, R.A., Lekic, V., Jackson, M.G., Kurz, M.D. (2020). Anomalous 182W in high 3He/4He ocean island basalts: Fingerprints of Earth's core? *Geochim. Cosmochim. Acta*, 271, 194–211.

Nakagawa, T. and Tackley, P.J. (2014). Influence of combined primordial layering and recycled morb on the coupled thermal evolution of Earth's mantle and core. *Geochem. Geophys. Geosyst.*, 15, 619–633.

Nicolaysen, K., Frey, F.A., Hodges, K.V., Weis, D., Giret, A. (2000). 40Ar/39Ar geochronology of food basalts from the Kerguelen Archipelago, southern Indian Ocean: Implications for Cenozoic eruption rates of the Kerguelen plume. *Earth Planet. Sci. Lett.*, 174, 313–328.

Olson, P., Schubert, G., Anderson, C. (1993). Structure of axisymmetric mantle plumes. *J. Geophys. Res.*, 98, 6829–6844.

Pearce, C.R., Cohen, A.S., Coe, A.L., Burton, K.W. (2008). Molybdenum isotope evidence for global ocean anoxia coupled with perturbations to the carbon cycle during the Early Jurassic. *Geology*, 36, 231–234.

Percival, L.M.E., Witt, M.L.I., Mather, T.A., Hermoso, M., Jenkyns, H.C., Hesselbo, S.P., Al-Suwaidi, A.H., Storm, M.S., Xu, W., Ruhl, M. (2015). Globally enhanced mercury deposition during the end-Pliensbachian extinction and Toarcian OAE: A link to the Karoo-Ferrar large igneous province. *Earth Planet. Sci. Lett.*, 428, 267–280.

Petterson, M.G. (2004). The geology of north and central Malaita, Solomon Islands: The thickest and most accessible part of the world's largest (Ontong Java) ocean plateau. In *Origin and Evolution of the Ontong Java Plateau*, Fitton, J.G., Mahoney, J.J., Wallace, P.J., Saunders, A.D. (eds). Geological Society, London.

Putirka, K.D. (2005). Mantle potential temperatures at Hawaii, Iceland, and the mid-ocean ridge system, as inferred from olivine phenocrysts: Evidence for thermally driven mantle plumes. *Geoch. Geoph. Geosyst.*, 6(5). doi: 10.1029/2005GC000915.

Reichow, M.K., Saunders, A.D., White, R.V., Pringle, M.S., Al'Mukhamedov, A.I., Medvedev, A.I., Kirda, N.P. (2002). Ar-40/Ar-39 dates from the West Siberian Basin: Siberian flood basalt province doubled. *Science*, 296, 1846–1849.

Renne, P.R., Ernesto, M., Pacca, I.G., Coe, R.S., Glen, J.M., Prévot, M., Perrin, M. (1992). The age of Parana flood volcanism, rifting of Gondwanaland, and the Jurassic-Cretaceous boundary. *Science*, 258, 975–979.

Renne, P.R., Zhang, Z.C., Richards, M.A., Black, M.T., Basu, A.R. (1995). Synchrony and causal relations between Permian-Triassic boundary crises and Siberian flood volcanism. *Science*, 269, 1413–1416.

Renne, P.R., Deckart, K., Ernesto, M., Feraud, G., Piccirillo, E.M. (1996). Age of the Ponta Grossa dike swarm (Brazil), and implications to Parana flood volcanism. *Earth Planet. Sci. Lett.*, 144, 199–211.

Richards, M.A., Duncan, R.A., Courtillot, V.E. (1989). Flood basalts and hot-spot tracks: Plume heads and tails. *Science*, 246, 103–107.

Richards, M.A., Jones, D.L., Duncan, R.A., DePaolo, D.J. (1991). A mantle plume initiation model for the Wrangellia flood basalt and other oceanic plateaus. *Science*, 254, 263–267.

Ritsema, H.J., van Heijst, J.H., Woodhouse, J.H. (1999). Complex shear velocity structure beneath Africa and Iceland. *Science*, 286, 1925–1928.

Rizo, H., Andrault, D., Bennett, N.R., Humayun, M., Brandon, A., Vlastelic, I., Moine, B., Poirier, A., Bouhifd, M.A., Murphy, D.T. (2019). 182W evidence for core-mantle interaction in the source of mantle plumes. *Geochem. Perspect. Lett.*, 11, 6–11.

Romanowicz, B. and Gung, Y.C. (2002). Superplumes from the core-mantle boundary to the lithosphere: Implications for heat flux. *Science*, 296, 513–516.

Rost, S., Garnero, E.J., Williams, Q., Manga, M. (2005). Seismological constraints on a possible plume root at the core-mantle boundary. *Nature*, 435, 666–669.

Schilling, J.G. (1973). Iceland mantle plume: Geochemical evidence along Reykjanes Ridge. *Nature*, 242, 565–571.

Schulte, P., Alegret, L., Arenillas, I., Arz, J.A., Barton, P.J., Bown, P.R., Bralower, T.J., Christeson, G.L., Claeys, P., Cockell, C.S. et al. (2010). The Chicxulub asteroid impact and mass extinction at the Cretaceous-Paleogene boundary. *Science*, 327, 1214–1218.

Self, S., Widdowson, M., Thordarson, T., Jay, A.E. (2006). Volatile fluxes during flood basalt eruptions and potential effects on the global environment: A Deccan perspective. *Earth Planet. Sci. Lett.*, 248, 518–532.

Sinton, C.W., Duncan, R.A., Storey, M., Lewis, J., Estrada, J.J. (1998). An oceanic flood basalt province within the Caribbean plate. *Earth Planet. Sci. Lett.*, 155, 221–235.

Sleep, N.H. (1990). Hotspots and mantle plumes: Some phenomenology. *J. Geophys. Res.*, 95, 6715–6736.

Sobolev, S.V., Sobolev, A.V., Kuzmin, D.V., Krivolutskaya, N.A., Petrunin, A.G., Arndt, N.T., Radko, V.A., Vasiliev, Y.R. (2011). Linking mantle plumes, large igneous provinces and environmental catastrophes. *Nature*, 477, 312–316.

Sprain, C.J., Renne, P.R., Vanderkluysen, L., Pande, K., Self, S., Mittal, T. (2019). The eruptive tempo of Deccan volcanism in relation to the Cretaceous-Paleogene boundary. *Science*, 363, 866–870.

Stacy, F.D. and Loper, D.E. (1983). The thermal boundary layer interpretation of D" and its role as a plume source. *Phys. Earth Planet. Inter.*, 33, 45–55.

Stixrude, L. and Lithgow-Bertelloni, C. (2012). Geophysics of chemical heterogeneity in the mantle. *Annu. Rev. Earth Planet. Sci.*, 40, 569–595.

Storey, B.C. (1995). The role of mantle plumes in continental breakup: Case histories from Gondwanaland. *Nature*, 377, 301–308.

Svensen, H., Planke, S., Chevallier, L., Malthe-Sorenssen, A., Corfu, F., Jamtveit, B. (2007). Hydrothermal venting of greenhouse gases triggering Early Jurassic global warming. *Earth Planet. Sci. Lett.*, 256, 554–566.

Svensen, H., Planke, S., Polozov, A.G., Schmidbauer, N., Corfu, F., Podladchikov, Y.Y., Jamtveit, B. (2009). Siberian gas venting and the end-Permian environmental crisis. *Earth Planet. Sci. Lett.*, 277, 490–500.

Swisher III., C.C., Grajales-Nishimura, J.M., Montanari, A., Margolis, S.V., Claeys, P., Alvarez, W., Renne, P., Cedillo-Pardo, E., Maurrasse, F.J.M.R., Curtis, G.H., Smith, J., McWilliams, M.O. (1992). Coeval 40Ar/39Ar ages of 65.0 million years ago from Chicxulub crater meltrock and Cretaceous-Tertiary boundary tektites. *Science*, 257, 954–958.

Tackley, P.J. (1998). Three-dimensional simulations of mantle convection with a thermo-chemical basal boundary layer: D"? In *The Core-Mantle Boundary Region*, Gurnis, M.E.A. (ed.). AGU, Washington, D.C.

Taylor, B. (2006). The single largest oceanic plateau: Ontong Java-Manihiki-Hikurangi. *Earth Planet. Sci. Lett.*, 241, 372–380.

Thorne, M.S. and Garnero, E.J. (2004). Inferences on ultralow-velocity zone structure from a global analysis of SPdKS waves. *J. Geophys. Res.*, 109, B08301.

Torsvik, T.H., Smethurst, M.A., Burke, K., Steinberger, B. (2008). Long term stability in deep mantle structure: Evidence from the ~300 Ma Skagerrak-Centered Large Igneous Province (the SCLIP). *Earth Planet. Sci. Lett.*, 267, 444–452.

Tsekhmistrenko, M., Sigloch, K., Hosseini, K., Barruol, G. (2021). A tree of Indo-African mantle plumes imaged by seismic tomography. *Nature Geosciences*, 14, 612–619.

Wei, S.S., Shearer, P.M., Lithgow-Bertelloni, C., Stixrude, L., Tian, D. (2020). Oceanic plateau of the Hawaiian mantle plume head subducted to the uppermost lower mantle. *Science*, 370, 983–987.

Wessel, P. (1993). Observational constraints on models of the Hawaiian hot spot swell. *J Geophys Res.*, 98, 16095–16104.

Wignall, P.B. (2001). Large igneous provinces and mass extinctions. *Earth Sci. Rev.*, 53, 1–33.

Wignall, P.B. (2011). Lethal volcanism. *Nature*, 477, 285–286.

Xu, Y., He, B., Chung, S.-L., Menzies, M.A., Frey, F.A. (2004). Geologic, geochemical, and geophysical consequences of plume involvement in the Emeishan flood-basalt province. *Geology*, 32, 917–920.

Yoshino, T., Makino, Y., Suzuki, T., Hirata, T. (2020). Grain boundary diffusion of W in lower mantle phase with implications for isotopic heterogeneity in oceanic island basalts by core-mantle interactions. *Earth Planet. Sci. Lett.*, 530, 115887.

Zhang, C.-L., Li, Z.-X., Li, X.-H., Xu, Y.-G., Zhou, G., Ye, H.-M. (2010). A Permian large igneous province in Tarim and Central Asian orogenic belt, NW China: Results of a ca. 275 Ma mantle plume? *GSA Bulletin*, 122, 2020–2040.

Zhou, Z., Malpas, J., Song, X.Y., Robinson, P.T., Sun, M., Kennedy, A.K., Lesher, C.M., Keays, R.R. (2002). A temporal link between the Emeishan large igneous province (SW China) and the end-Guadalupian mass extinction. *Earth. Planet. Sci. Lett.*, 196, 113–122.

5

Heat Flow and Secular Cooling of the Mantle

Stéphane LABROSSE

LGL TPE, CNRS, Université de Lyon, ENS de Lyon, Université Lyon 1, France

5.1. Introduction

As stated by Sadi Carnot in the first page of his seminal book (Figure 5.1), heat transfer is at the origin of most of the dynamics of the Earth. The dynamics of the mantle is no exception: it can be described as a convecting layer. Quantifying the heat flow out of the Earth is therefore a requirement to know the energy available to drive mantle convection. Conversely, models of mantle convection and evolution must satisfy this simple constraint to be deemed acceptable. The goal of this chapter is therefore to provide an overview of the energy budget of the Earth and in particular its mantle. I also shortly discuss the implications of this energy budget for the thermal evolution of the Earth and its dynamics and readers interested by more details about the thermal evolution are invited to read the relevant chapter in volume 4 on the archean and proterozoic Earth. A more complete discussion of the topics in this chapter can be found in Jaupart et al. (2015), Labrosse (2016) and Jaupart and Mareschal (2011).

Structure and Dynamics of the Earth's Interior 1,
coordinated by Julien MONTEUX.

RÉFLEXIONS

SUR LA

PUISSANCE MOTRICE

DU FEU.

PERSONNE n'ignore que la chaleur peut être la cause du mouvement, qu'elle possède même une grande puissance motrice: les machines à vapeur, aujourd'hui si répandues, en sont une preuve parlante à tous les yeux.

C'est à la chaleur que doivent être attribués les grands mouvemens qui frappent nos regards sur la terre; c'est à elle que sont dues les agitations de l'atmosphère, l'ascension des nuages, la chute des pluies et des autres météores, les courans d'eau qui sillonnent la surface du globe et dont l'homme est parvenu à employer pour son usage une faible partie; enfin les tremblemens de terre, les éruptions volcaniques, reconnaissent aussi pour cause la chaleur.

Figure 5.1. *Sadi Carnot at age 17 (left) and the first page of his book (right). For a color version of this figure, see www.iste.co.uk/monteux/earth1.zip*

5.2. Geophysics of the seafloor and the oceanic heat flow

The ocean floor having been out of reach to scientists for a long time, ideas about deformations of the solid mantle first came from observations on continents, that is, parallelism of the coasts of Africa and South-America, correspondence of geologic outcrops, etc. These observations were compiled by Wegener in his famous book on continental drift first published in 1915. Several authors proposed the possibility of mantle convection in the early 20th century (Holmes 1931; Pekeris 1935) but systematic exploration of the seafloor was the turning point that led to the development of plate tectonics and the general acceptance of mantle convection as its engine. The current theories and models of mantle convection also focus largely on the oceanic part of the Earth surface, which exhibits the clearest signs of mantle dynamics. The simple interpretation of these observations in terms of plate tectonics can be used to predict the thermal structure of the oceanic plates which, by comparisons to observations, permits an accurate determination of the total oceanic heat flow.

5.2.1. *Infinite half-space model*

Let us first consider a simple model schematic of an oceanic plate (Figure 5.2).

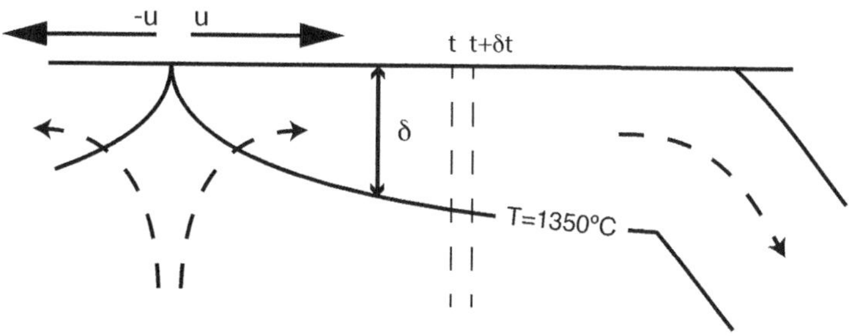

Figure 5.2. *Schematic model of the oceanic lithosphere produced at a ridge and moving away from it at a velocity u. Hot matter moves up under the ridge and away horizontally at the surface where it cools by contact with the oceanic water. A cooling front is illustrated by a temperature contour $T = 1.350\,°\mathrm{C}$*

In this first model, the oceanic plate is considered as a two-dimensional object, whose structure is independent from the along-ridge direction. Subduction pulls the oceanic plate, which opens the ridge and, by mass conservation, pulls up mantle material toward the surface. This hot matter cools down by diffusion in contact with oceanic water, whose temperature is kept roughly constant by efficient convection, about $2\,°\mathrm{C}$. This cooling process occurs as the solid material moves away from the ridge, which is represented in Figure 5.2 by deepening of the temperature contour $T = 1350\,°\mathrm{C}$ (the choice of this value will be justified below). Taking $\kappa = 10^{-6}\,\mathrm{m}^2\,\mathrm{s}^{-1}$ as thermal diffusivity, the depth affected by this cooling after a time t is $\delta = \sqrt{\kappa t}$, which reaches $\delta_{max} = 75\,\mathrm{km}$ for the maximum age of the oceanic lithosphere, $180\,\mathrm{Myr}$. This value can be used to make several simplifications to the complete problem. First, δ_{max} is much smaller than the Earth radius whose curvature can therefore be neglected for this problem. It is also much larger than the typical topography difference between ridges and abyssal plains of order $3\,\mathrm{km}$. Both these observations lead to the infinite-half-space model, that is, considering a flat, horizontally infinite upper boundary. δ_{max} is also small compared to the typical size of oceanic plates, several thousands of kilometers. This means that in Figure 5.2, the horizontal scale is in fact much larger than the vertical one. Therefore, we expect that the contour line is close to being horizontal and we can approximate the temperature gradient as being vertical. With that approximation, the equation satisfied by the temperature T *as we follow the plate* is

$$\frac{\partial T}{\partial t} = \kappa \frac{\partial^2 T}{\partial z^2}, \qquad [5.1]$$

which is simply the classical one-dimensional diffusion equation. This equation neglects radiogenic heating, which is reasonable since the oceanic mantle is rather depleted in incompatible elements (Jochum et al. 1983) so that its integration over the limited depth of interest gives a negligible contribution.

Solving equation [5.1] requires the specification of boundary and initial conditions. At the surface, $z = 0$, we impose the constant temperature of the ocean, which we set to 0 °C for simplicity. At the ridge, we assume the temperature to be independent of depth, $T = T_M$. This problem has a famous solution, first derived by Fourier (1822) and used by Kelvin (Thomson 1862) to compute the age of the Earth, which is

$$T = T_M \operatorname{erf}\left(\frac{z}{2\sqrt{\kappa t}}\right), \qquad [5.2]$$

with z the depth below the surface and erf the error function. Taking the partial derivative with respect to depth and multiplying by the thermal conductivity k gives, by Fourier's law, the heat flux density at the surface as a function of the seafloor age:

$$-q = k \frac{\partial T}{\partial z}(0, t) = \frac{k T_M}{\sqrt{\pi \kappa t}}. \qquad [5.3]$$

We consider here $-q$ which is positive, because it is the component of the flux vector toward the decreasing z, that is, upward, since we have considered z to be the depth. We have therefore obtained an expression of the oceanic heat flow density expressed as a function of its age and we can now test this expression by comparison with observations.

5.2.2. *Observations: seafloor age and heat flow*

In order to test the prediction from the infinite half-space model, we need measurements of heat flux density on places of ocean floor that are well dated.

The age of the seafloor can be obtained using measurements of magnetic anomalies combined with the geomagnetic polarity reversal timescale (Cande and Kent 1995). Figure 5.3 shows a map obtained with this approach (Müller et al. 2008).

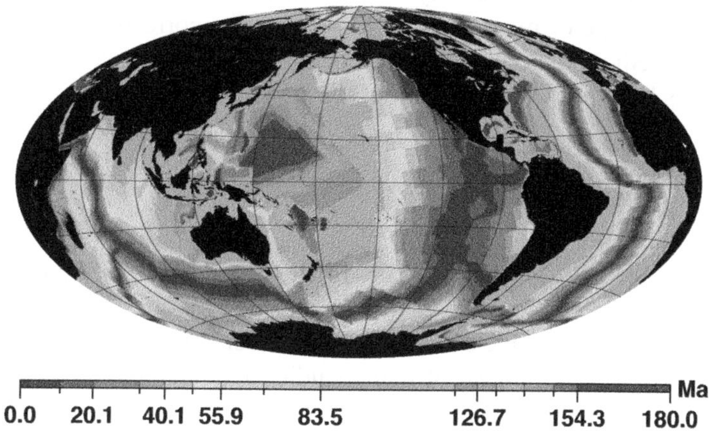

Figure 5.3. *Map of the seafloor age obtained using magnetic anomalies. The age model is from Müller et al. (2008) and can be retrieved from http://www.earthbyte.org/Resources/agegrid2008.html. For a color version of this figure, see www.iste.co.uk/monteux/earth1.zip*

In principle, measuring the heat flow should be easy. From Fourier's law, we see that we need to measure the temperature gradient and the thermal conductivity and multiply the two. Several difficulties however arise. First, this must be done at the bottom of the oceans, several kilometers under the surface. To accomplish that, specific probes have been developed. A several-meters long needle-like stick under a heavy weight mass is brought down to the ocean floor and left to penetrate under its own weight in unconsolidated sediments. The probe is equipped with regularly spaced temperature probes. This allows to readily measure the temperature gradient. Measuring the thermal conductivity is more complex and several approaches have been used. The first approach was to recover core samples and measure it in the lab, but this is very time consuming. Also, the conductivity of sediments happens to vary significantly on small distances for sediments, owing to their very heterogeneous water content. A much better approach is to heat up the needle and measure the response in terms of temperature, which depends directly on the thermal conductivity (Von Herzen and Maxwell 1959). This approach is much more efficient and also more reliable since the temperature gradient and the thermal conductivity are measured at the same place.

Figure 5.4 shows a compilation of data from well-sedimented area obtained with in situ thermal conductivity measurements (Lister et al. 1990). Each box gives the range of measurements from a region corresponding to an age range. The prediction from the infinite half-space model is compared to the observations by fitting a relationship of the form

$$q = \frac{C_Q}{\sqrt{t}}, \qquad [5.4]$$

with $C_Q = 490 \pm 20$ mW/m^2 Myr$^{1/2}$. Figure 5.4 shows that the theory of a half-space cooling gives a very good fit to the data for ages lower than about 80 Myr but that a systematic deviation can be observed for older seafloor. This deviation has been the subject of many studies and a simple explanation can be the onset of small-scale convection below the lithosphere (e.g. Davaille and Jaupart 1994). Indeed, convective heat transfer being more efficient than conduction, we expect that small-scale convection makes the heat flow larger than it would be in the case of the conductive cooling of an infinite half space.

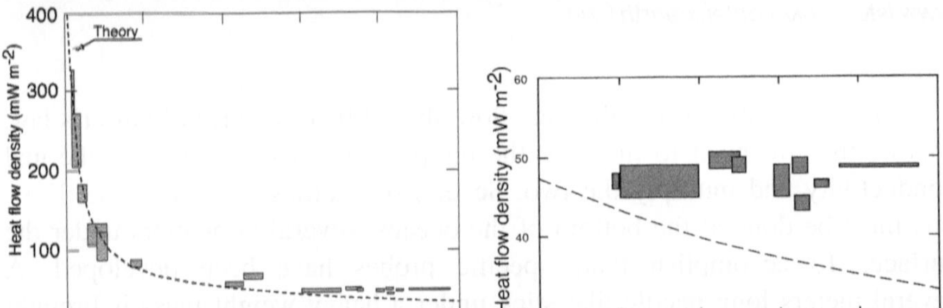

Figure 5.4. *Compilation of good quality oceanic heat flow data, that is, obtained in well-sedimented areas with in situ thermal conductivity measurement. Each box gives the variability of measurements in each age bin. The dashed line is a fit using equation [5.4]. The right subfigure is a zoom for large seafloor ages showing a systematic deviation of the theory from observations. Adapted from Lister et al. (1990). For a color version of this figure, see www.iste.co.uk/monteux/earth1.zip*

From the value of C_Q, and measurements on the thermal conductivity and diffusivity of lithospheric rocks, one can estimate the temperature of the mantle below the lithosphere (potential temperature) as $T_M \simeq 1350\,°C$. This is compatible with the melting temperature of basalts as determined by geochemistry and petrology (Kinzler and Grove 1992). This fit has also been tested against high-resolution heat flow data obtained near the Juan de Fuca ridge by Davis et al. (1999) (see Figure 5.5). The agreement between the fit obtained on the global database and the high-resolution observations on the young seafloor is remarkable when the sedimentary cover is thick enough to prevent hydrothermal circulation. On the other hand, at a distance to the ridge lower than $40\,\mathrm{km}$, the sedimentary cover is too thin and the hydrothermal circulation through it makes the conductive heat flow drop below the predicted value. In this area, heat is transferred by porous flow through the sediments. This illustrates an additional difficulty in measuring the oceanic heat flow: using Fourier's law relies on the assumption that heat transfer across the sediments happens only by conduction, which breaks down when the sediment cover is not sufficient.

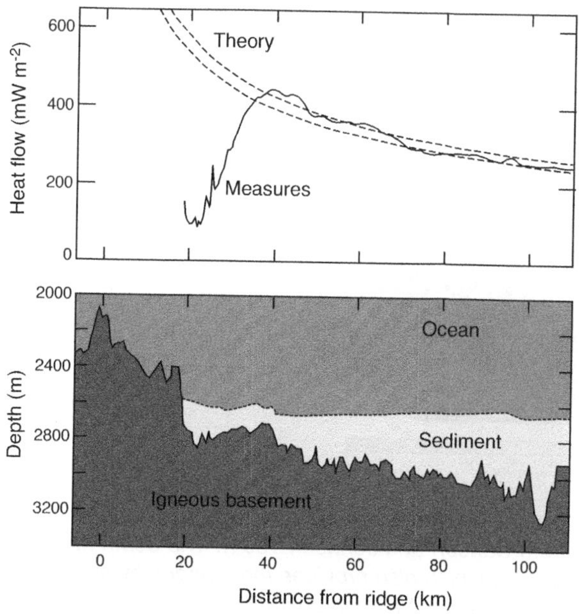

Figure 5.5. *Heat flow density (top) and structure of the seafloor near the Juan de Fuca ridge (bottom), as a function of the distance to the ridge. The dashed lines correspond to prediction of heat flow using equation [5.4]. Adapted from Davis et al. (1999) and Jaupart et al. (2015). For a color version of this figure, see www.iste.co.uk/monteux/earth1.zip*

5.2.3. *Seafloor depth*

Another piece of information about the seafloor comes from the observation of its depth. Figure 5.6 shows a global map of the seafloor depth. It can be combined with the map of seafloor age to represent the depth as a function of age. This was done by Carlson and Johnson (1994) and Figure 5.6 shows the result of this analysis. It shows clearly that, for ages less than 90 Myr, the depth of the seafloor increases linearly with the square-root of its age, an observation that must be related to its thermal structure. As temperature decreases, most materials experience contraction, that is, an increase in density ρ. This is in fact the origin of thermal convection when this happens in a gravity field: the material that cools at the top of the mantle gets denser than the underlying one and drives subduction. This phenomenon of thermal contraction is quantified by a thermodynamical parameter, the thermal expansion coefficient,

$$\alpha = -\frac{1}{\rho}\frac{\partial \rho}{\partial T}. \qquad [5.5]$$

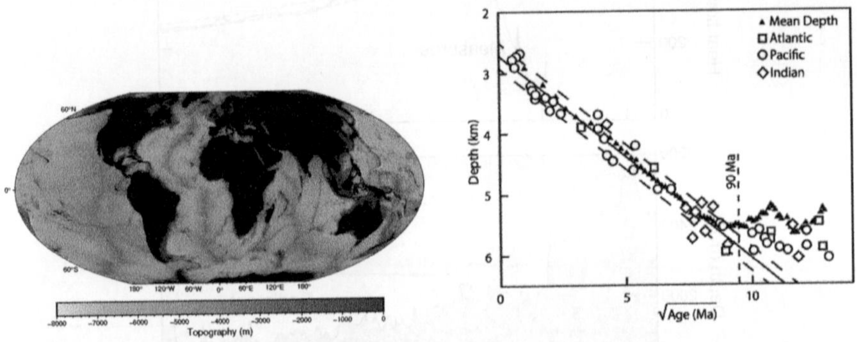

Figure 5.6. *Map of the seafloor depth (left) and depth as a function of lithospheric age (right, adapted from Carlson and Johnson 1994). The depth map is plotted using the Generic Mapping Tool, which also provides the topography model. For a color version of this figure, see www.iste.co.uk/ monteux/earth1.zip*

Using T_M as reference temperature for which the density is ρ_M and assuming that α is constant, we can write a simple equation of state for the plate material as

$$\rho = \rho_M \left[1 - \alpha \left(T - T_M\right)\right]. \qquad [5.6]$$

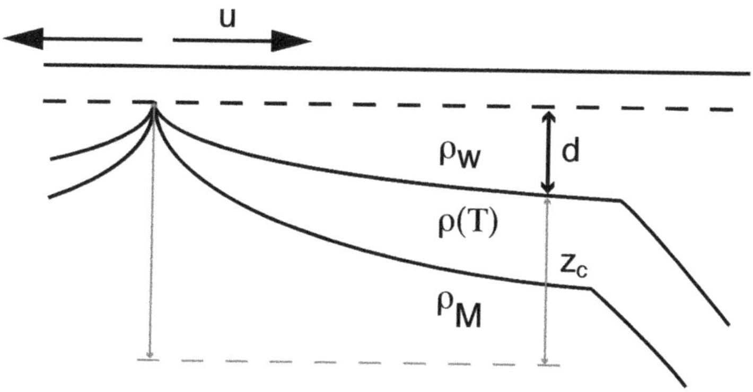

Figure 5.7. *Schematic model of the oceanic lithosphere with exaggerated topography. The horizontal level depicted by the red dashed line is called the isostatic compensation level. For a color version of this figure, see www.iste.co.uk/monteux/earth1.zip*

As the lithosphere cools down, it contracts, its density increases, which deforms the underlying asthenospheric mantle. We assume that isostasy holds: there is a depth in the mantle where pressure is laterally uniform. This pressure is dominated by the hydrostatic part and this means that the total weight down to this depth is laterally uniform. We note $d(t)$ the difference in seafloor depth between age t and age 0 (Figure 5.7). ρ_w is the density of water. The uniform pressure at compensation depth means that

$$\rho_M(d + z_c) = \rho_w d + \int_0^{z_c} \rho dz \Rightarrow d = \frac{1}{\rho_M - \rho_w} \int_0^{z_c} (\rho - \rho_M) dz. \qquad [5.7]$$

The density difference between the deep mantle (ρ_M) and that of the lithosphere (ρ) depends on temperature and is written using equation [5.6]:

$$\begin{aligned} d &= \frac{\rho_M \alpha}{\rho_M - \rho_w} \int_0^{z_c} (T_M - T)\, dz, \\ &= \frac{\rho_M \alpha T_M}{\rho_M - \rho_w} \int_0^{z_c} \left[1 - \operatorname{erf}\left(\frac{z}{2\sqrt{\kappa a}}\right)\right] dz, \\ &= \frac{2\rho_M \alpha T_M \sqrt{\kappa a}}{\rho_M - \rho_w} \int_0^{z_c/2\sqrt{\kappa a}} [1 - \operatorname{erf} u]\, du, \end{aligned} \qquad [5.8]$$

where a change of variable was performed to get the last expression. Since the variation of temperature occurs only at small depth, in the lithosphere, changes of the integral with z_c are negligible as soon as $z_c \gg \sqrt{\kappa t}$. We can then compute the integral using $z_c \to \infty$, knowing that

$$\int_0^\infty (1 - \operatorname{erf} x) dx = \frac{1}{\sqrt{\pi}}, \quad [5.9]$$

and we get

$$d = \frac{2\rho_M \alpha T_M}{\rho_M - \rho_w} \sqrt{\frac{\kappa t}{\pi}}. \quad [5.10]$$

This expression provides a good explanation of the deepening of the ocean floor linearly with the square root of its age. The deviation for large ages can again be explained by the onset of small-scale convection below the lithosphere (Davaille and Jaupart 1994). Adjusting this expression to the observations provides an estimate of the mantle temperature that is independent from the one obtained using the heat flow data. The two estimates are compatible, which confirms the validity of the model.

5.2.4. *Total oceanic heat loss*

As shown in the previous subsections, several independent data sets are compatible with the infinite-half-space cooling model. This model provides a prediction of the oceanic heat flow as a function of lithospheric age. Using the fit to the observations presented in Figure 5.4, the seafloor age map (Figure 5.3) can be used to create a map of oceanic heat flow for normal oceans (Figure 5.8). To be precise, the heat flow is computed with

$$q = \begin{cases} \frac{490}{\sqrt{t}} & \text{if } 1\,\text{Myr} < t \leq 80\,\text{Myr}, \\ \frac{490}{\sqrt{80}} & \text{if } t > 80\,\text{Myr} \end{cases} \quad [5.11]$$

with q the heat flow density in $\mathrm{mW\,m^{-2}}$ and t the age of seafloor in million years. The expression is in fact truncated at age less than 1 Myr to cope with the divergence toward infinity at zero age. Clearly, the assumptions behind the infinite-half-space model break down at the ridge where heat is mostly transferred by hydrothermal convection and magma transport, which makes it nearly impossible to measure with precision. When this occurs, the conductive heat flux drops but the total (conductive and convective) oceanic heat flow is expected to follow the same evolution with time. Davis et al. (1999) showed that, close to the Juan de Fuca ridge, the infinite-half-space law that was established for large seafloor ages (Figure 5.4) still applies as close as $40\,\mathrm{km}$ away from the ridge axis, where a heat flux density of $400\,\mathrm{mW\,m^{-2}}$ is observed. This is also confirmed by the observations of the seafloor depth near ridges.

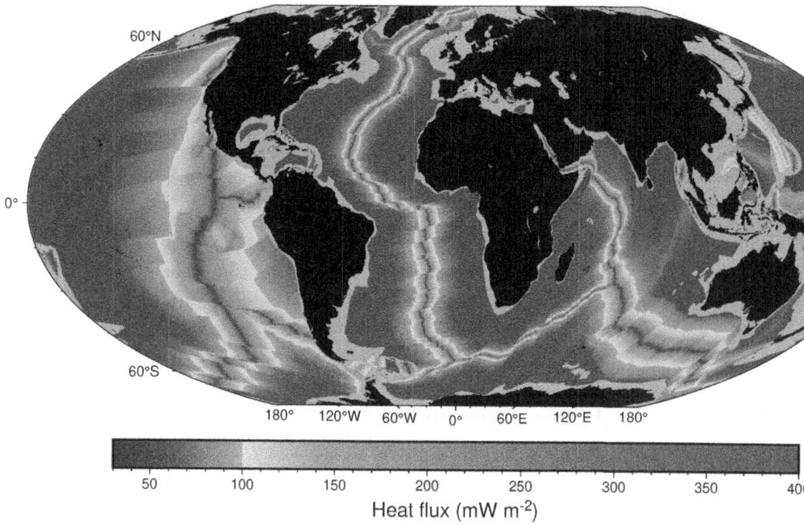

Figure 5.8. *Oceanic heat flow computed from the seafloor age map. For a color version of this figure, see www.iste.co.uk/monteux/earth1.zip*

From the map if Figure 5.3, it is possible to compute the distribution of seafloor ages (Sclater et al. 1980; Cogné and Humler 2004), which represents the non-normalized probability density of each age as a function of age. In other words, $\frac{\mathrm{d}A}{\mathrm{d}t}(t)\,\mathrm{d}t$ is the area of seafloor that has an age comprised between t and $t+\mathrm{d}t$. The total seafloor area is A_{oc} and the maximum age observed is $t_{max} = 180\,\mathrm{Myr}$. This distribution can be used to compute the total heat flow as

$$Q_{oc} = \int_{A_{oc}} q\,\mathrm{d}A = \int_0^{t_{max}} q(t)\,\frac{\mathrm{d}A}{\mathrm{d}t}\,(t)\,\mathrm{d}t. \qquad [5.12]$$

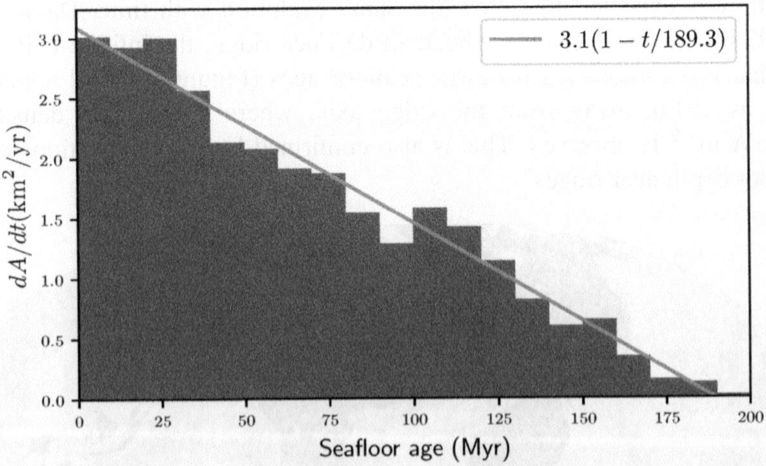

Figure 5.9. *Distribution of seafloor ages. For a color version of this figure, see www.iste.co.uk/monteux/earth1.zip*

The seafloor age distribution corresponding to the map shown in Figure 5.3 is plotted in Figure 5.9. It has been observed to be vary grossly linearly with age (Parsons 1982; Cogné and Humler 2004),

$$\frac{\mathrm{d}A}{\mathrm{d}t} = C_0\left(1 - \frac{t}{t_{max}}\right), \qquad [5.13]$$

with $C_0 = 3.1\,\mathrm{km}^2\,\mathrm{yr}^{-1}$ the present day accretion rate at ridges. This parameter is linked to the maximum age by the total oceanic area as

$$A_{oc} = \int_0^{t_{max}} \frac{\mathrm{d}A}{\mathrm{d}t}(t)\,\mathrm{d}t = \frac{C_0 t_{max}}{2}. \qquad [5.14]$$

Injecting these two relationships in equation [5.12], and using the expression of the heat flow as a function of the seafloor age (equation [5.3]), we get the total oceanic heat loss as

$$Q_{oc} = \frac{4kT_M}{3\sqrt{\pi\kappa}} C_0 \sqrt{t_{max}} = \frac{8}{3} A_{oc} \frac{kT_M}{\sqrt{\pi\kappa t_{max}}}. \qquad [5.15]$$

In order to get a numerical value, we can estimate the different parameters from laboratory measurements on samples collected in the field. There is always the issue of representativity of the samples to consider and the simplest is in fact to use the fit of observations by the model. This amounts to assuming that $kT_M/\sqrt{\pi\kappa} = 490\,\mathrm{W\,m^{-2}\,Ma^{1/2}}$ and we get $Q_{oc} = 29\,\mathrm{TW} = 29 \times 10^{12}\,\mathrm{W}$, taking into account the constant value of the flux for ages larger than 80 Myr (Jaupart et al. 2015).

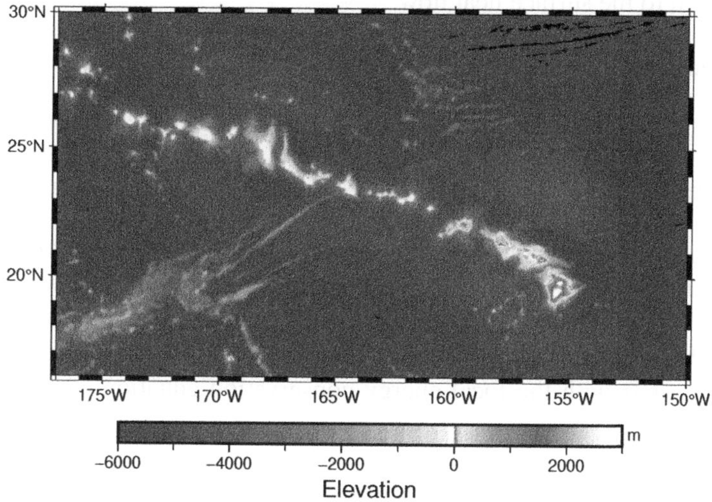

Figure 5.10. *Topography around the Hawaiian island chain. For a color version of this figure, see www.iste.co.uk/monteux/earth1.zip*

This value of the oceanic heat flow does not account for regions of intraplate volcanism, often referred to as hotspots (see Chapter 4). These regions are characterized by the presence of volcanic islands or seamounts and a larger scale topographic swell surrounding the island chain (Figure 5.10). It also produces a well-defined geoid anomaly. Morgan (1971) proposed that these island chains are caused by convective plumes originating from the deep mantle that impinge under the plate and maintaining the observed swell. Several authors (Davies 1988; Sleep 1990; Ribe and Christensen 1999) have developed simple dynamical models for this process and shown that the swell amplitude and the geoid can be used to estimate the buoyancy flux of the plume. Considering a thermal origin to this buoyancy, this can be converted into a net contributions of mantle plumes to the surface heat flow that has to be added to that from the normal ocean floor. One difficulty with this exercise comes largely from identifying which among the regions of intraplate volcanism are associated with a mantle plume (Clouard and Bonneville 2001). However, the largest contribution comes from hotspots whose plume origin is usually accepted like Hawaii, Réunion or Iceland. With this approach, a total contribution of intraplate volcanism to the surface heat flow in the range $2\,\mathrm{TW}$ to $4\,\mathrm{TW}$ is generally accepted. Even though the surface expression of mantle plumes can be impressive, in particular in terms of volcanism, their quite reduced area makes them provide a modest contribution to the surface heat flow.

Combining the heat flow from normal oceans to the contribution from intraplate volcanism gives a total ocean heat flow of $Q_{oc} = 32\,\mathrm{TW}$.

5.2.5. *Link to mantle dynamics*

As explained above, the oceanic heat flow is measured using Fourier's law of heat conduction. The oceanic lithosphere being the thermal boundary layer of mantle convection, the total conductive heat flow at the surface should be matched by the total convective heat flow below the lithosphere. To show that, let us start with the local energy conservation equation (see Chapter 1 for details):

$$\rho C \left(\frac{\mathrm{d}T}{\mathrm{d}t} + \boldsymbol{v} \cdot \boldsymbol{\nabla} T \right) = \boldsymbol{\nabla} \cdot (k \boldsymbol{\nabla} T) + \rho h \qquad [5.16]$$

where T is the temperature, $\boldsymbol{v}$ is the flow velocity vector, ρ is the density, C is the heat capacity, k is the thermal conductivity and h is the heat production. Assuming for simplicity that ρ and C are constant, this equation can be integrated over the volume $V(r)$ contained between the sphere $A(r)$ of radius r and the Earth surface $r = R_e$ to express the surface heat flow Q_s as a function of its contribution in the spherical shell of volume $V(r)$:

$$\begin{aligned} Q_s \equiv - \int_{A(R_e)} k \frac{\partial T}{\partial r} \mathrm{d}A = & - \int_{V(r)} \rho C \frac{\partial T}{\partial t} \mathrm{d}V + \int_{A(r)} \rho C v_r T \, \mathrm{d}A \\ & - \int_{A(r)} k \frac{\partial T}{\partial r} \mathrm{d}A + \int_{V(r)} \rho h \, \mathrm{d}V, \end{aligned} \qquad [5.17]$$

where the non-penetration boundary condition at the Earth surface, $v_r(R_e) = 0$, has been used along with the divergence theorem and the mass conservation equation for an incompressible fluid (Boussinesq approximation), $\boldsymbol{\nabla} \cdot \boldsymbol{v} = 0$. We have, by order of appearance on the right-hand side, the secular cooling of the spherical shell, the convective heat flow across $A(r)$, the conductive heat flow across $A(r)$ and the radiogenic heating of the shell. Mass conservation, again using the incompressibility assumption, implies that the radial velocity v_r has a zero mean across $A(r)$:

$$\overline{v_r} \equiv \frac{1}{A(r)} \int_{A(r)} v_r \,\mathrm{d}A = 0. \qquad [5.18]$$

Without loss of generality, the temperature can be split into its horizontal average and deviations from it, $T = \overline{T} + \delta T$. The convective heat flow across $A(r)$ can then be written as

$$\begin{aligned} Q_{conv}(r) = \textstyle\int_{A(r)} \rho C v_r T \,\mathrm{d}A &= \rho C \overline{T} \textstyle\int_{A(r)} v_r \,\mathrm{d}A + \textstyle\int_{A(r)} \rho C v_r \delta T \,\mathrm{d}A \\ &= A(r) \rho C \overline{v_r \delta T}. \end{aligned} \qquad [5.19]$$

This shows that the convective heat flow is associated with the correlation between v_r and δT, two fields with a zero mean whose product must have a non-zero mean for convective heat transfer to proceed. Physically, cold material ($\delta T < 0$) is more dense than the average and sinks ($v_r < 0$), which contributes positively to outward heat transfer. Similarly, hotter than average material ($\delta T > 0$) leads to upward flow ($v_r > 0$), which also contribute positively.

Equation [5.17] can then be written differently to make the horizontal mean of the temperature appear explicitly:

$$\begin{aligned} Q_s \equiv -A(R_e) k \frac{\partial \overline{T}}{\partial r}(R_e) = A(r) \rho C \overline{v_r \delta T} - A(r) k \frac{\partial \overline{T}}{\partial r} \\ + \textstyle\int_{V(r)} \rho h \,\mathrm{d}V - \textstyle\int_{V(r)} \rho C \frac{\partial T}{\partial t} \,\mathrm{d}V. \end{aligned} \qquad [5.20]$$

Figure 5.11 shows a snapshot of a convective model in a spherical shell with a constant internal heating rate (left) and time-averaged minimum, average and maximum temperature profiles along with the profiles of conductive and convective heat flux (right). The heat flux profiles are scaled by a constant spherical surface. All quantities are here dimensionless and only the general behavior is to be discussed here. The calculation has been done until a statistically steady state has been reached, so that averaging over a sufficiently long time allows us to remove the secular evolution term on the right-hand side of equation [5.20]. Efficient convection tends to make the temperature uniform in the bulk of the spherical shell, as can be seen in the mean temperature profile, although lateral variations remain important as they drive flow and heat transfer. It means that the conduction heat transfer (second term on the right-hand side of equation [5.20]) is negligible in the bulk of the shell, which is confirmed on the right panel of Figure 5.11.

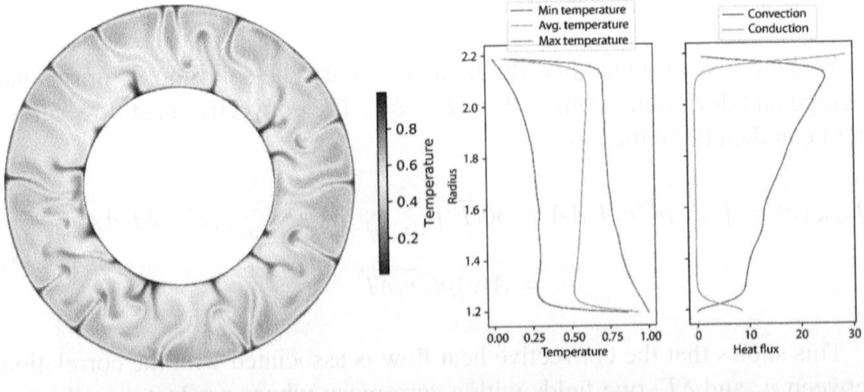

Figure 5.11. *Rayleigh-Bénard convection in a spherical shell with internal heating. Left: a snapshot of the temperature field. Middle: temporal mean of minimum, average and maximum temperature profiles. Right: temporal mean of the convection (advection) and diffusion heat transfer profiles. The calculation has been done using the mantle convection code StagYY with the spherical annulus geometry (Tackley 2008; Hernlund and Tackley 2008). For a color version of this figure, see www.iste.co.uk/monteux/earth1.zip*

On the other hand, convective heat transfer is important in the bulk of the domain, but must go to zero at the horizontal boundaries where both v_r and δT are null. Close to the boundaries, conductive heat transfer must take over, which requires a large mean temperature gradient and creates the boundary layers that can be seen in Figure 5.11. As we go deeper in the mantle, the energy balance equation [5.20] includes a larger volume in which radiogenic heat is produced. Since the total surface heat flow is independent from the volume over which we consider the balance, this means that the convective heat flow must decrease with depth, which is observed in Figure 5.11. Similar profiles are discussed in Chapter 1.

Applying equation [5.17] to the oceanic part of the mantle, assuming now for simplicity that continents are perfect insulators over the mantle (a more complete view is discussed in the next section), we can consider the sphere $A(r)$ just underneath the lithosphere. By so doing, we can follow the discussion of the previous section and neglect both the secular cooling of the lithosphere and its radiogenic heat production. Indeed, the volume involved is too small as well as the concentration of heat producing elements (Jochum et al. 1983) for this contribution to matter and we expect that the surface heat flow must be approximately matched by the convective heat transfer below the lithosphere. Using a simple schematic model of a subducting oceanic plate (Figure 5.12), we can estimate that heat transfer. Using equation [5.19] for the convective heat flow and assuming that the flow velocity is uniform in the subducting plate, we need to estimate its mean temperature anomaly. The oceanic plate thickens by diffusion and eventually gets back down at a subduction zone. Just after turning at the subduction, we can assume that little diffusion has had time to proceed and the thermal structure of the plate is largely inherited from what it was just before subduction. The BB' temperature profile across the subducting plate is then a rotated version of the AA' profile in the oceanic lithosphere, although it heats up and widens by contact with the mantle wedge. The integrated temperature anomaly per unit length of subduction is then simply the area of the blue triangle in Figure 5.12, that is, $-T_M\delta/2$. Integrating over the whole surface of subduction gives a total subducting heat flow:

$$Q_{sub} = -\rho C v_r \frac{T_M \delta}{2} L_s \quad [5.21]$$

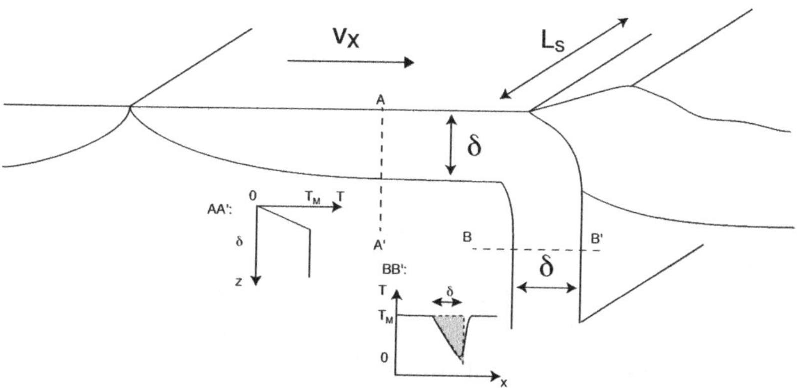

Figure 5.12. *Schematic model of a subducting oceanic plate. For a color version of this figure, see www.iste.co.uk/monteux/earth1.zip*

where v_r is the mean vertical velocity of the subducting plates and $L_s \sim 5 \times 10^4$ km is the total length of subduction. Taking $\rho = 3 \times 10^3\,\text{kg}\,\text{m}^{-3}$,

$C = 10^3\,\mathrm{J\,kg^{-1}\,K^{-1}}$, $T_M = 1300\,°\mathrm{C}$ (note that T_M is in fact the temperature difference to the ocean bottom one), $\delta = -100\,\mathrm{km}$ and $v_r = -10\,\mathrm{cm\,yr^{-1}}$ we get $Q_{sub} = 32\,\mathrm{TW}$.

A similar calculation can be done for a hot plume impinging under an oceanic plate. The first difference with the calculation just performed is that the vertical velocity and the temperature anomaly are positive instead of both being negative. However, as shown by equation [5.19], in both cases the result provides a positive contribution to the total heat flow. The second, major, difference is in the values of the parameters entering the calculation: for a plume, the temperature anomaly is of the order $\delta T = 200\,\mathrm{K}$ and the associated surface is $A = \pi R_p^2$ with $R_p \simeq 100\,\mathrm{km}$ the plume radius. This leads to $Q_p = 0.06\,\mathrm{TW}$ for a single plume. Assuming that 30 of such plumes exist leads to a total contribution of about 2 TW, of the same order of magnitude as that obtained from the study of hotspot swells (Davies 1988).

Although the near perfect match to the total oceanic heat flow could be seen as fortuitous considering the gross nature of the calculation, it shows that heat transport below the lithosphere is largely dominated by down-welling cold plates and that upwelling currents play a limited role in comparison.

5.3. Continental heat flow

The study of continental heat flow presents challenges that are different from those for the oceans. Working on continents can be seen as easier than in the deep oceans. However, while the deep ocean temperature is nearly constant, the temperature at the surface of continents shows a large variability with time on daily to climatic timescales. These fluctuations penetrate into the Earth and the determination of the heat flow requires to measure the temperature gradient and thermal conductivity at depth larger than that affected by these fluctuations. This can be determined by solving the skin-effect problem of thermal diffusion in an infinite-half-space with periodically varying surface temperature. The solution of this problem is well known and shows temperature fluctuating at each depth with an amplitude decaying exponentially with depth. Without solving completely the mathematical problem, it is possible to predict the penetration depth of a time-fluctuating surface temperature using a dimensional analysis. The variation with time of the surface temperature can be decomposed in Fourier modes and, the diffusion equation being linear, the effect of each mode can be considered independently from the others. Considering a time dependence of the surface temperature in the form $\exp(i2\pi t/\tau)$, τ being the period of interest, the infinite half-space thermal diffusion problem provides only one possible natural length scale, $\delta = \sqrt{\kappa\tau}$. A full treatment of this problem indeed leads to the well-known skin-effect solution: a downward propagating wave that is attenuated with depth z as $\exp(-z\sqrt{\pi/\kappa\tau})$. Therefore, the fluctuation of the surface temperature is attenuated by a factor $1/e$ when the skin-depth $z_\kappa = \sqrt{\kappa\tau/\pi}$ is reached. Taking $\kappa = 10^{-6}\,\mathrm{m}^2\,\mathrm{s}^{-1}$ as typical value for the thermal diffusivity, we get $z_\kappa = 17\,\mathrm{cm}$ for a daily variation ($\tau = 1\,\mathrm{d}$), $z_\kappa = 3\,\mathrm{m}$ for a yearly variation, $\tau = 1\,\mathrm{yr}$, and $z_\kappa = 460\,\mathrm{m}$ for the shortest Milankovitch cycle for which $\tau = 21\,\mathrm{kyr}$. This means that the continental crust records at depth the variations of the surface temperature and borehole temperature measurements can be used to determine past surface temperatures (e.g. Mareschal and Beltrami 1992). The downside is that very deep boreholes are needed to determine the heat flow coming from the Earth interior. Fortunately, the amplitude of the climatic variation, which penetrates the deepest, is much smaller than that of daily and yearly variations and, in practice, boreholes deeper than $100\,\mathrm{m}$ provide useful results (see e.g. Jaupart and Mareschal 1999).

The temperature is measured as a function of depth using cables to bring down sensors and the thermal conductivity of rocks is measured on core samples. The energy balance of a given borehole can be used to assess whether the heat flux measured is representative of that coming from depth: as for the oceanic case discussed above, when considering a thin layer of the crust (typically less than $2\,\mathrm{km}$ deep), the volumetric terms are normally negligible, except for that from heat capacity close to the surface. In that case, we expect that, below the upper part that is affected by surface processes, the heat flux is independent from depth and that value is representative of the geothermal flux. Figure 5.13 shows several temperature profiles from various boreholes in the Canadian shield. They all show a negative gradient with depth close to the surface, associated with the heating up of the surface since the last deglaciation. It means that the Earth interior is currently receiving heat from the atmosphere in this area. The situation reverses after a few tens of meters and the data deeper than about $100\,\mathrm{m}$ can be used to obtain the heat flux. Note also that some profiles show a kink at some depth, for example, the one for Gull River at about $440\,\mathrm{m}$. At that location, the temperature gradient changes from about $12\,\mathrm{K\,km}^{-1}$ to about $15\,\mathrm{K\,km}^{-1}$. This is associated with a change of thermal conductivity from $2.8\,\mathrm{W\,m}^{-1}\,\mathrm{K}^{-1}$ to $2.2\,\mathrm{W\,m}^{-1}\,\mathrm{K}^{-1}$, which gives a consistent heat flux value of $35\,\mathrm{mW\,m}^{-2}$ (Perry et al. 2004).

Collecting data from all continents allows one to produce a map of the continental heat flux density, which can be combined with that for the oceans discussed in the previous section. This gives a global heat flux map shown in Figure 5.14 from the model of Jaupart and Mareschal (2011). An interesting point to note is that the continental heat flux is generally low, particularly in cratons, similar to that of deep oceans. However, tectonically active regions can reach large values, for example, in the western Americas.

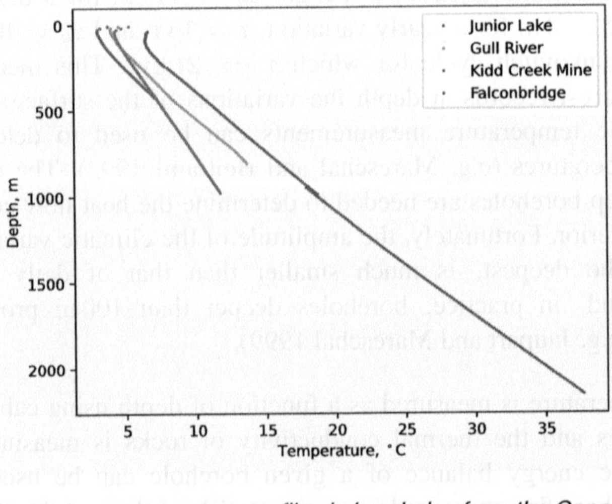

Figure 5.13. *Various temperature profiles in boreholes from the Canadian shield. Data from Perry et al. (2004, 2006, 2009). For a color version of this figure, see www.iste.co.uk/monteux/earth1.zip*

The large short-scale (~500 km) variations of the surface heat flux in continents can be used to infer constraints on the thermal structure of continents (Mareschal and Jaupart 2004). Because diffusion tends to erase lateral variations of temperature, the lateral variations observed at the surface must be explained by variations of larger amplitude at depth, with an amplification factor that depends exponentially with z/L_x, z being depth and L_x the horizontal wavelength. The observations in Canada clearly imply that lateral variations of the surface heat flow must take their origin in the shallow crust, from lateral variation of radiogenic heating, not the mantle (Mareschal and Jaupart 2004). These observations can then be used to estimate the heat flux from the mantle in stable continental areas, which is found to be in the range $11\,\mathrm{W\,m^{-1}\,K^{-1}}$ to $18\,\mathrm{W\,m^{-1}\,K^{-1}}$. Combining the heat flow measurements from all continents gives a total continental heat flow $Q_{co} = 14\,\mathrm{TW}$ (see Jaupart et al. 2015, for a more complete discussion).

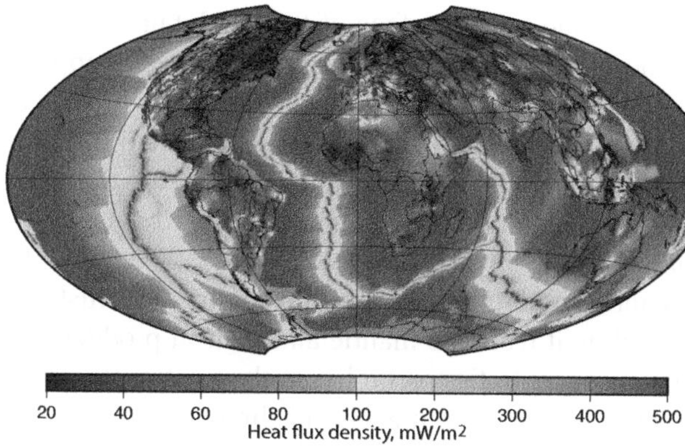

Figure 5.14. *Global heat flow map. The model is from Jaupart and Mareschal (2011). The oceanic part is computed from the seafloor age and the continental part from a global heat flow database. For a color version of this figure, see www.iste.co.uk/monteux/earth1.zip*

5.4. Heat sources and secular evolution

Having discussed the heat loss, we can now turn to heat sources and then discuss the secular evolution of the mantle. The two main heat input to the mantle come from radioactivity and from the core. Other sources, for example, associated with tidal deformation, are clearly negligible (Jaupart et al. 2015).

Several radioactive isotopes are responsible for heating up terrestrial planets. They can be separated into two groups: short-lived and long-lived ones. The short-lived isotopes have a half-life much shorter than the age of the Solar System and are now completely extinct. The two relevant isotopes are ^{60}Fe, which decays to ^{60}Ni with a half-life $\tau_{1/2} = 1.49\,\text{Myr}$, and ^{26}Al, which decays to ^{26}Mg with a half-life $\tau_{1/2} = 717\,\text{kyr}$. These two isotopes are important for explaining the melting and differentiation of planetesimals and planetary embryos (Šrámek et al. 2012) but are long dead at the time of formation of the Moon (Touboul et al. 2007), which marks the end of the Earth accretion for the most part. Their lasting role lies in their effect on the thermal and compositional structure of the forming bricks of the Earth, and therefore its initial condition, but they need not be considered when dealing with the recent evolution of the Earth.

The long-lived isotopes are those that still exist and produce heat today: ^{238}U ($\tau_{1/2} = 4.46\,\text{Gyr}$), ^{235}U ($\tau_{1/2} = 0.704\,\text{Gyr}$), ^{232}Th ($\tau_{1/2} = 14\,\text{Gyr}$) and ^{40}K ($\tau_{1/2} = 1.26\,\text{Gyr}$). The concentration in these isotopes can be readily measured on rocks found at the surface of the Earth but determining the bulk composition of the mantle is much more complex. Several approaches can be used combining sampling from different rocks at the surface and measurements from meteorites, which are left-overs from the accretion of the Earth. The result depends on the assumptions done for the calculation (see the discussion by Jaupart et al. 2015), which leads to a 15% uncertainty on the total heat production, which at present is $H_{tot} = 20 \pm 3\,\text{TW}$ for the whole silicate Earth (mantle and crust). Because the continental crust is produced by extraction of partial melt from the mantle and all heat producing elements are incompatible, their concentration is higher in the continental crust than in the mantle. Using heat flux measurements and direct sampling, the radiogenic heat production rate in the continental crust is estimated to be $H_{cc} = 7 \pm 1\,\text{TW}$, that is, one-third of the total (Jaupart et al. 2015).

Having determined the concentration in each heat producing isotope, it is straightforward to compute the time evolution of the radiogenic heating as the sum of the contributions from each isotope, each decaying exponentially with its own half-life. The result of this calculation is shown in Figure 5.15 and we see that the total radiogenic heating has decreased by a factor of nearly 5 since the formation of the Earth.

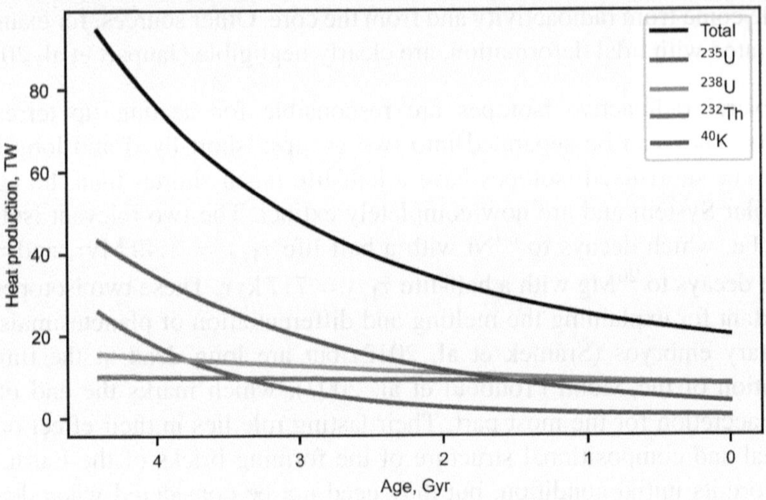

Figure 5.15. *Radiogenic heat production in the mantle as a function of time. Contributions from the four long-lived heat producing isotopes are shown in color, and the total in black. For a color version of this figure, see www.iste.co.uk/monteux/earth1.zip*

The other major source of heat for the mantle is the heat flow from the core. The value of that heat flow is quite uncertain and arguments to constrain it come from the thermodynamics of the Earth core sustaining the magnetic field by dynamo action and from seismological observations of the lowermost mantle and their interpretations.

Because the outer core is a low viscosity liquid, it cannot sustain large lateral temperature variations. The secular variations of the magnetic field, which is associated with flow in the outer core, can be used to estimate these lateral variations and they are of the order 10^{-4} K (Labrosse et al. 1997). These are completely negligible compared to lateral variations of temperature in the mantle, of the order 10^3 K, for example, associated with subducting plates (section 5.2.5). Therefore, the mantle sees the core as a uniform temperature hot bath. In terms of physical model of mantle convection, it means that the relevant boundary condition to apply for the mantle at its bottom is that of a fixed temperature. In return, it implies that the heat flux that the mantle imposes to the core is laterally variable. The situation is similar to what is observed at the surface of the mantle with strong lateral variations of heat flux density at the bottom of the ocean (Figure 5.8) whose temperature is nearly uniform. The discussion of the heat flow from the core must therefore address both the total (or averaged) value and its lateral variations. The former is more relevant than the latter to thermal evolution issues but seismological observations of the lowermost mantle are by nature local, and not complete, which complicates their interpretations.

The first evidence that the CMB heat flow is important comes from the existence of mantle plumes and their surface expressions, the hotspots (see Chapter 4). However, it is important to realize that the regions of large heat flow at the core mantle boundary are associated with the arrival of cold material, not the departure of hot plumes (Labrosse 2002). The situation is symmetrical to what can be observed at the surface where the large heat flow is associated with ridge while the downward motion of cold material (subduction) is associated with a low heat flow at the surface. The heat flow at the CMB could be estimated using a global circulation model of the mantle (e.g. Coltice et al. 2019) but the results from such models strongly depend on several poorly constrained ingredients, like the core-mantle boundary temperature or the presence of compositional variations in the lowermost mantle.

The other potential constraint on the thermal structure of the CMB region comes from seismology. The presence and detection of the bridgmanite to post-perovskite phase transition (Murakami et al. 2004) offers the hope of direct constraints on temperature in the lowermost mantle. Hernlund et al. (2005) showed that the observations of double seismic discontinuities in some regions of the D" region and of no discontinuity in others could be explained by the curvature of the temperature profile as a function of height above the CMB and different mantle temperatures above it. In cold regions, associated with the arrival of subducting plates, the authors expect the presence of a lens of post-perovskite sandwiched between bridgmanite above and below. On the other hand, no bridgmanite is expected in hot regions that are at the origin of hot plumes. The situation is in fact more complex since the phase equilibrium also depends on composition and a post-perovskite lens has been proposed to be present at the bottom of the mantle in the central Pacific in a low seismic velocity region that is interpreted as hot and FeO-rich (Lay et al. 2006).

Seismic tomography could also, in principle, permit the direct calculation of the CMB heat flux. The idea is simple: seismic velocity depends on temperature which can then be computed from a tomography models. The heat flux is then obtained as the product of the temperature gradient by thermal conductivity. Many difficulty arise in this approach because each step is prone to several sources of uncertainty. For example, the seismic velocities depend also on composition, not only temperature. The use of P wave velocity reduces that specific problem and Hernlund and McNamara (2015) discuss in details how the various seismological observations can be combined to obtain a map of the CMB heat flux. They conclude that the total heat flow across the CMB is 13 ± 3 TW.

This important parameter can also be addressed by considering the dynamics and evolution of the core. The magnetic field of the Earth is produced by dynamo action in the liquid core, driven by convective flow (see the review by Landeau et al. 2022). At the present time, the buoyancy source sustaining that process is primarily compositional (Labrosse 2015): the cooling of the core makes the inner core grow by crystallization, which tends to enrich the lowermost outer core in the light elements that are depleted in the solid inner core. However, the inner core has a small mass and growing it to its present size requires to extract only about 30×10^{28} J from the core (Labrosse et al. 2001; Labrosse 2015). With a nominal CMB heat flow of 10 TW, the inner grows to its present size in 1 Gyr. The Earth magnetic field has existed for at least 3.5 Gyr (Tarduno et al. 2010), which means that either the inner core is older than that, requiring a very small CMB heat flow, or dynamo action was sustained differently before the inner core nucleation. The simplest idea is that of thermal convection but it requires a CMB heat flow larger than the one conducted down the isentropic gradient at the top of the core, which is an uncertain but potentially large value. Indeed, the thermal conductivity of the core is debated, with values proposed in the range $20\,\mathrm{W\,m^{-1}\,K^{-1}}$ to $250\,\mathrm{W\,m^{-1}\,K^{-1}}$ (see Landeau et al. 2022, and references therein). With the highest values, it would be difficult to maintain dynamo action by thermal convection alone, and a value of $100\,\mathrm{W\,m^{-1}\,K^{-1}}$ requires a CMB heat flow larger than about $15\,\mathrm{TW}$ to maintain thermal convection. Alternatively, convection could be maintained by extracting light elements from the top of the core (O'Rourke and Stevenson 2016; Badro et al. 2016; Hirose et al. 2017) but this scenario requires further inquiries to be confirmed. It is worth noting that the minimum heat flow to maintain dynamo action by thermal convection, albeit still poorly constrained, is similar to the value that can be estimated from studies of the mantle structure discussed in the previous paragraph.

Having identified heat sources and sinks for the mantle, we can now discuss its present cooling rate. The global energy balance for the mantle can be written as

$$MC\frac{\mathrm{d}T_p}{\mathrm{d}t} = H + Q_{CMB} - Q_s, \qquad [5.22]$$

where $M = 4\times10^{24}\,\mathrm{kg}$ is the mass of the silicate Earth and $C \simeq 1200\,\mathrm{J\,kg^{-1}\,K^{-1}}$ is the heat capacity of the mantle. Combining the continental and oceanic areas, the total heat flow from the solid Earth is estimated to be $Q_s = 46 \pm 3\,\mathrm{TW}$ (Jaupart et al. 2015). Considering the central values for the radiogenic heat production, $H = 20\,\mathrm{TW}$ and the CMB heat flow $Q_{CMB} = 13\,\mathrm{TW}$ gives a present cooling rate for the mantle of $85\,\mathrm{K\,Gyr^{-1}}$. This central value is obviously quite uncertain but matches the observations from the petrology of non-arc basalts (Herzberg et al. 2010). Alternatively, the CMB heat flow being uncertain, assuming the same heat capacity for the whole Earth gives a mean present cooling rate of $114\,\mathrm{K\,Gyr^{-1}}$.

Extrapolating that value backward in time requires a thermal evolution model. Presenting this problem in detail goes beyond the scope of the present chapter. Here only the general ideas in that topic are presented.

Equation [5.22] can be used to compute the thermal evolution of the Earth if the surface and CMB heat flows are available as a function of time. In principle, the surface heat flow could be obtained using plate tectonics reconstruction (Becker et al. 2009) but this is limited to the last few 100 Myr of the Earth history. To go beyond that, we need a model for heat transfer by mantle convection, typically using scaling laws relating the surface heat flux to the Rayleigh number. This approach has been used for decades but it was soon recognized that standard scaling laws led to a thermal catastrophe paradox: computing the thermal evolution backward from the present time leads to a fully molten Earth about 1 Gyr ago (Christensen 1985), at odds with all observations. The reason for this is simple: the present day surface heat flow is much larger than the radiogenic heating rate. Going backward in time implies a larger temperature in the past and therefore a smaller viscosity and an even larger surface heat flow. Several approaches have been proposed to solve that problem, starting with a radiogenic heating rate larger than what geochemical studies imply, unconventional heat flux scaling laws which cut the feedback loop of the temperature-dependent viscosity, or, more reasonable, a CMB heat flow larger than originally envisioned (Driscoll and Bercovici 2014; Labrosse 2016; Patočka et al. 2020). A CMB heat flow of order 15 TW indeed acts as a heat source for the mantle and avoids the mantle viscosity to decrease too fast going backward. On the other hand, it implies that the core cools down faster than the mantle and therefore was much hotter in the past. Since the present day CMB temperature is presently close to the solidus of mantle materials at that pressure (Andrault et al. 2011; Fiquet et al. 2010) and the ultra-low velocity zones (ULVZ) observed by seismology are most readily explained by the presence of partial melt (Williams and Garnero 1996), it implies that the lowermost mantle must have been largely molten in the past. This led to the scenario of the basal magma ocean (Labrosse et al. 2007) in which we proposed that a thick layer of dense FeO-rich melt slowly crystallized with time below the solid mantle. The last remnants would be the partially molten pockets seen as ULVZs. The presence of such a basal magma ocean is expected to greatly affect convection in the solid (Labrosse et al. 2018; Agrusta et al. 2019) and the implications in terms of thermal evolution are still to be fully explored (see Chapter 3).

5.5. Conclusion

The present day heat budget of the Earth is rather well constrained, with a surface heat flow amounting to 46 TW and a radiogenic heating rate of about 20 TW. These two numbers can be related to what is known of the dynamics of the mantle. They can also be used to compute the present-day average cooling rate of the Earth, 120 K Gyr^{-1} (Jaupart et al. 2015). Difficulties arise when we want to break that number into the core and mantle parts, which requires knowledge of the heat flow at the CMB, and when we want to extrapolate that number toward the early Earth. Both these issues are still unresolved but a large CMB heat flow, $Q_{CMB} \geq 13\,\mathrm{TW}$, permits us to solve the long-standing problem of the thermal evolution of the Earth, provided the existence and long-lasting basal magma ocean is recognized. Many issues remain to be addressed and looking at this problem using the thermal energy budget is a very powerful test.

5.6. References

Agrusta, R., Morison, A., Labrosse, S., Deguen, R., Alboussière, T., Tackley, P.J., Dubuffet, F. (2019). Mantle convection interacting with magma oceans. *Geophys. J. Int.*, 220, 1878–1892.

Andrault, D., Bolfan-Casanova, N., Nigro, G.L., Bouhifd, M.A., Garbarino, G., Mezouar, M. (2011). Solidus and liquidus profiles of chondritic mantle: Implication for melting of the Earth across its history. *Earth Planet. Sci. Lett.*, 304, 251–259.

Badro, J., Siebert, J., Nimmo, F. (2016). An early geodynamo driven by exsolution of mantle components from Earth's core. *Nature*, 536, 326–328.

Becker, T.W., Conrad, C.P., Buffett, B.A., Müller, R.D. (2009). Past and present seafloor age distributions and the temporal evolution of plate tectonic heat transport. *Earth Planet. Sci. Lett.*, 278, 233–242.

Cande, S.C. and Kent, D.V. (1995). Revised calibration of the geomagnetic polarity timescale for the late cretaceous and cenozoic. *J. Geophys. Res.*, 100(B4), 6093–6095.

Carlson, R.L. and Johnson, H.P. (1994). On modeling the thermal evolution of the oceanic upper-mantle: An assessment of the cooling plate model. *J. Geophys. Res.*, 99(B2), 3201–3214.

Christensen, U.R. (1985). Thermal evolution models for the Earth. *J. Geophys. Res.*, 90, 2995–3007.

Clouard, V. and Bonneville, A. (2001). How many Pacific hotspots are fed by deep mantle plumes? *Geology*, 29(8), 695–698.

Cogné, J.-P. and Humler, E. (2004). Temporal variation of oceanic spreading and crustal production rates during the last 180 My. *Earth Planet. Sci. Lett.*, 227, 427–439.

Coltice, N., Husson, L., Faccenna, C., Arnould, M. (2019). What drives tectonic plates? *Sci. Adv.*, 5(10), eaax4295.

Davaille, A. and Jaupart, C. (1994). Onset of thermal convection in fluids with temperature-dependent viscosity: Application to the oceanic mantle. *J. Geophys. Res.*, 99, 19853–19866.

Davies, G.F. (1988). Ocean bathymetry and mantle convection: 1. Large-scale flow and hotspots. *J. Geophys. Res.*, 93, 10467–10480.

Davis, E.E., Chapman, D.S., Wand, K., Villinger, H., Fisher, A.T., Robinson, S.W., Grigel, J., Pribnow, D., Stein, J., Becker, K. (1999). Regional heat flow variations across the sedimented Juan de Fuca ridge eastern flank: Constraints on lithospheric cooling and lateral hydrothermal heat transport. *J. Geophys. Res.*, 104, 17,675–17,688.

Driscoll, P. and Bercovici, D. (2014). On the thermal and magnetic histories of Earth and Venus: Influences of melting, radioactivity, and conductivity. *Phys. Earth Planet. Inter.*, 236, 36–51.

Fiquet, G., Auzende, A.L., Siebert, J., Corgne, A., Bureau, H., Ozawa, H., Garbarino, G. (2010). Melting of peridotite to 140 GigaPascals. *Science*, 329, 1516–1518.

Fourier, J.B.J. (1822). *Théorie analytique de la chaleur*. Didot, Paris.

Hernlund, J.W. and McNamara, A. (2015). The core–mantle boundary region. In *Treatise on Geophysics*, 2nd edition, Schubert, G. (ed.). Elsevier, Oxford.

Hernlund, J.W. and Tackley, P.J. (2008). Modeling mantle convection in the spherical annulus. *Phys. Earth Planet. Inter.*, 171(1–4), 48–54.

Hernlund, J.W., Thomas, C., Tackley, P.J. (2005). Phase boundary double crossing and the structure of Earth's deep mantle. *Nature*, 434, 882–886. doi: 10.1038/nature03472.

Herzberg, C., Condie, K., Korenaga, J. (2010). Thermal history of the Earth and its petrological expression. *Earth Planet. Sci. Lett.*, 292(1–2), 79–88.

Hirose, K., Morard, G., Sinmyo, R., Umemoto, K., Hernlund, J.W., Helffrich, G., Labrosse, S. (2017). Crystallization of silicon dioxide and compositional evolution of the Earth's core. *Nature*, 543, 7643.

Holmes, A. (1931). Radioactivity and Earth movements. *Trans. Geol. Soc. Glasgow*, 18, 559–606.

Jaupart, C. and Mareschal, J.C. (1999). The thermal structure and thickness of continental roots. *Lithos*, 48, 93–114.

Jaupart, C. and Mareschal, J.C. (2011). *Heat Generation and Transport in the Earth*. Cambridge University Press, Cambridge.

Jaupart, C., Labrosse, S., Lucazeau, F., Mareschal, J.-C. (2015). Temperatures, heat, and energy in the mantle of the earth. In *Treatise on Geophysics*, 2nd edition, Schubert, G. (ed.). Elsevier, Oxford.

Jochum, K.P., Hoffmann, A.W., Ito, E., Seufert, H.M., White, W.M. (1983). K, U and Th in mid-ocean ridge basalt glasses and heat production. K/U and K/Rb in the mantle. *Nature*, 306, 431–436.

Kinzler, R.J. and Grove, T.L. (1992). Primary magmas of mid-ocean ridge basalts 2. Applications. *J. Geophys. Res.*, 97(16), 6907–6926.

Labrosse, S. (2002). Hotspots, mantle plumes and core heat loss. *Earth Planet. Sci. Lett.*, 199(1–2), 147–156 [Online]. Available at: https://www.sciencedirect.com/science/article/pii/S0012821X0200537X.

Labrosse, S. (2015). Thermal evolution of the core with a high thermal conductivity. *Phys. Earth Planet. Inter.*, 247, 36–55.

Labrosse, S. (2016). Thermal state and evolution of the earth core and deep mantle. In *Deep Earth: Physics and Chemistry of the Lower Mantle and Core*, Terasaki, H., Fischer, R. (eds). AGU Publications, Washington D.C.

Labrosse, S., Poirier, J.-P., Le Mouël, J.-L. (1997). On cooling of the Earth's core. *Phys. Earth Planet. Inter.*, 99, 1–17.

Labrosse, S., Poirier, J.-P., Le Mouël, J.-L. (2001). The age of the inner core. *Earth Planet. Sci. Lett.*, 190, 111–123.

Labrosse, S., Hernlund, J.W., Coltice, N. (2007). A crystallizing dense magma ocean at the base of Earth's mantle. *Nature*, 450, 866–869.

Labrosse, S., Morison, A., Deguen, R., Alboussière, T. (2018). Rayleigh-Bénard convection in a creeping solid with a phase change at either or both horizontal boundaries. *J. Fluid Mech.*, 846, 5–36.

Landeau, M., Fournier, A., Nataf, H.C., Cébron, D., Schaeffer, N. (2022). Sustaining Earth's magnetic dynamo. *Nature Reviews Earth and Environment*, 3(4), 255–269.

Lay, T., Hernlund, J., Garnero, E.J., Thorne, M.S. (2006). A post-perovskite lens and D" heat flux beneath the central Pacific. *Science*, 314(5803), 1272–1276 [Online]. Available at: http://www.sciencemag.org/cgi/content/abstract/314/5803/1272.

Lister, C.R.B., Sclater, J.G., Davis, E.E., Villinger, H., Nagihara, S. (1990). Heat flow maintained in ocean basins of great age: investigations in the north-equatorial west pacific. *GJI*, 102(3), 603–628.

Mareschal, J.-C. and Beltrami, H. (1992). Evidence for recent warming from perturbed geothermal gradients: Examples from eastern Canada. *Climate Dynamics*, 6, 135–143.

Mareschal, J.-C. and Jaupart, C. (2004). Variations of surface heat flow and lithospheric thermal structure beneath the North American craton. *Earth Planet. Sci. Lett.*, 223(1–2), 65–77.

Morgan, W.J. (1971). Convection plumes in the lower mantle. *Nature*, 230(5288), 42–43.

Müller, R.D., Sdrolias, M., Gaina, C., Roest, W.R. (2008). Age, spreading rates, and spreading asymmetry of the world's ocean crust. *G3*, 9.

Murakami, M., Hirose, K., Kawamura, K., Sata, N., Ohishi, Y. (2004). Post-perovskite phase transition in $MgSiO_3$. *Science*, 304, 855–858.

O'Rourke, J.G. and Stevenson, D.J. (2016). Powering earth's dynamo with magnesium precipitation from the core. *Nature*, 529(7586), 387–389.

Parsons, B. (1982). Causes and consequences of the relation between area and age of the ocean floor. *J. Geophys. Res.*, 87, 289–302.

Patočka, V., Šrámek, O., Tosi, N. (2020). Minimum heat flow from the core and thermal evolution of the Earth. *Physics of the Earth and Planetary Interiors*, 305(March), 106457. doi: 10.1016/j.pepi.2020.106457.

Pekeris, C.L. (1935). Thermal convection in the interior of the Earth. *Geophys. J. R. Astr. Soc.*, 3(8), 343–367.

Perry, H.K., Jaupart, C., Mareschal, J.C., Rolandone, F., Bienfait, G. (2004). Heat flow in the Nipigon arm of the Keweenawan rift, northwestern Ontario, Canada. *Geophys. Res. Lett.*, 31(15), 1–5.

Perry, H.K., Jaupart, C., Mareschal, J.C., Bienfait, G. (2006). Crustal heat production in the superior province Canadian shield, and in North America inferred from heat flow data. *J. Geophys. Res.*, 111(4), 1–20.

Perry, H.K., Mareschal, J.C., Jaupart, C. (2009). Enhanced crustal geo-neutrino production near the Sudbury Neutrino Observatory, Ontario, Canada. *Earth Planet. Sci. Lett.*, 288(1–2), 301–308. doi: 10.1016/j.epsl.2009.09.033.

Ribe, N.M. and Christensen, U.R. (1999). The dynamical origin of Hawaiian volcanism. *Earth Planet. Sci. Lett.*, 171(4), 517–531.

Sclater, J.G., Jaupart, C., Galson, D. (1980). The heat flow through oceanic and continental crust and the heat loss of the Earth. *Rev. Geophys. Space Phys.*, 18, 269–312.

Sleep, N.H. (1990). Hotspots and mantle plumes: Some phenomenology. *J. Geophys. Res.*, 95, 6715–6736.

Šrámek, O., Milelli, L., Ricard, Y., Labrosse, S. (2012). Thermal evolution and differentiation of planetesimals and planetary embryos. *Icarus*, 217(1), 339–354.

Tackley, P.J. (2008). Modelling compressible mantle convection with large viscosity contrasts in a three-dimensional spherical shell using the yin-yang grid. *Phys. Earth Planet. Inter.*, 171(1–4), 7–18.

Tarduno, J.A., Cottrell, R.D., Watkeys, M.K., Hofmann, A., Doubrovine, P.V., Mamajek, E.E., Liu, D., Sibeck, D.G., Neukirch, L.P., Usui, Y. (2010). Geodynamo, solar wind, and magnetopause 3.4 to 3.45 billion years ago. *Science*, 327(5970), 1238–1240.

Thomson, W.L.K. (1862). On the secular cooling of the earth. *Trans. Roy. Soc. Edinburgh*, 23, 295–311.

Touboul, M., Kleine, T., Bourdon, B., Palme, H., Wieler, R. (2007). Late formation and prolonged differentiation of the moon inferred from W isotopes in lunar metals. *Nature*, 450(7173), 1206–1209.

Von Herzen, R. and Maxwell, A.E. (1959). The measurement of thermal conductivity of deep-sea sediments by a needle-probe method. *J. Geophys. Res.*, 64(10), 1557–1563.

Williams, Q. and Garnero, E.J. (1996). Seismic evidence for partial melt at the base of Earth's mantle. *Science*, 273(5281), 1528–1530.

6

Noble Gases: Geochemical Tracers of Mantle Dynamics

Manuel Alexis MOREIRA
ISTO, BRGM, CNRS, Université d'Orléans, France

6.1. Introduction

A fundamental question in Earth sciences and planetology is the origin of the atmosphere and oceans and what this means for Earth's habitability. The atmosphere is a fragile geological reservoir that is difficult to understand in terms of its origin and evolution. This requires an understanding of its interaction with the Earth's mantle, both for degassing and for the recycling of volatile elements in the mantle, in addition to the problem of dissipation from planetary atmospheres. The present and past dynamics of the Earth's mantle thus have a strong impact on the origin and evolution of our atmosphere.

The use of noble gases as tracers of mantle/atmosphere history began in the late 1960s and early 1970s with the pioneering work of Soviet researchers on ^{3}He of mantle origin (Mamyrin et al. 1969a, 1969b; Tolstikhin et al. 1974) and American researchers on xenon (Reynolds 1960; Butler et al. 1963). This discipline then took off in the 1970s and 1980s with a majority of studies clearly demonstrating the mantle origin of certain noble gases in samples of ridge or hotspot lavas, as well as CO_2-rich gases

Structure and Dynamics of the Earth's Interior 1,
coordinated by Julien MONTEUX.

(Butler et al. 1963; Hennecke and Manuel 1975; Craig and Lupton 1976, 1981; Kaneoka and Takaoka 1977; Phinney et al. 1978; Kurz and Jenkins 1981; Kurz et al. 1982a, 1982b, 1983; Staudacher and Allègre 1982; Poreda and Radicati di Brozolo 1984; Sarda et al. 1988).

Improved measurement and sampling techniques, combined with an increase in the number of samples analyzed, have made it possible to establish a systematic approach to noble gas geochemistry in the mantle. This gives a fairly clear picture of the origin and evolution of these and other volatile elements on Earth, from the accretion stage of our planet to the processes linked to its current dynamics.

In this chapter, section 6.2 covers current knowledge of noble gas composition in the Earth's mantle, with particular reference to its heterogeneity. The evolution of the mantle–atmosphere pair is then discussed in section 6.3 through the study of degassing and regassing of the Earth's mantle. The dynamics of the present and early mantle as seen through the lens of noble gases will be presented in section 6.4. Finally, in section 6.5, three open questions on the geochemistry of noble gases on Earth will be addressed.

6.2. Noble gases in the mantle

6.2.1. *General information on noble gases*

Noble gases are the chemical elements in the last column of the periodic table. Their valence layer is saturated, consequently they do not form covalent bonds with other atoms. They are therefore chemically inert. There are six of them, forming a homogeneous chemical family of monoatomic gases. From lightest to heaviest, they include helium (He), neon (Ne), argon (Ar), krypton (Kr), xenon (Xe), and radon (Rn). The latter has no stable isotope and will not be discussed in this chapter. These are highly volatile elements (Figure 6.1) with very low condensation temperatures (<3 K for helium, 9.3 K for neon, 48 K for argon, 53 K for krypton and 69 K for xenon; Lodders 2003). This explains why their abundances are very low in solids (chondrites/planets) compared with solar abundances, especially for helium and neon.

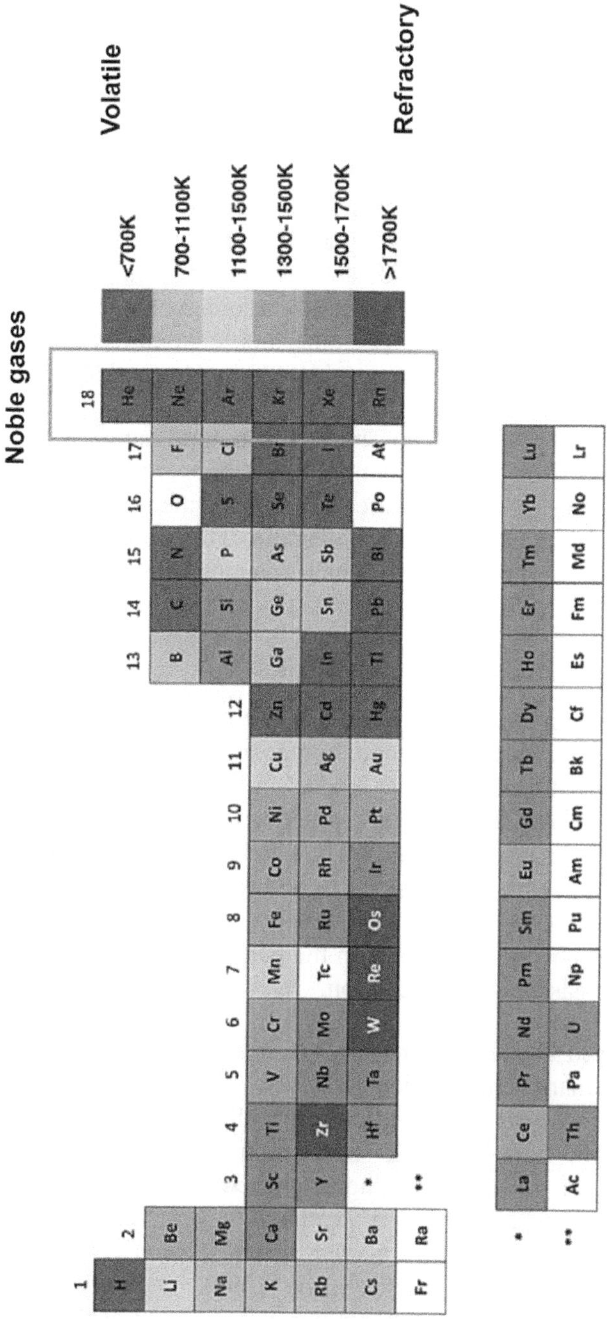

Figure 6.1. *Periodic table of the elements. For a color version of this figure, see www.iste.co.uk/monteux/earth1.zip*

COMMENT ON FIGURE 6.1.– *Column 18 contains noble gases (rare gases). These elements are highly volatile during condensation processes. They are chemically inert and therefore cannot be stably incorporated into minerals. They are very rare in Solar System solids, despite their high stellar abundance.*

Taking helium as an example, its abundance in the upper mantle is ~1 ppb, even though it is the second most abundant element in the Solar System. This figure is even overestimated, as it is mainly radiogenic helium (^{4}He/alpha particle) from the radioactive decay chains of U and Th. The amount of early ^{4}He is therefore more in the order of a few tens of ppt and falls in the ultra-trace category. The other noble gases have similar or even lower concentrations and are also considered ultra-trace elements on Earth.

As well as being difficult to analyze in natural samples, they are also sensitive to atmospheric contamination. Their monoatomic, inert and volatile nature also makes them sensitive to thermal metamorphism, and their preservation in ancient rocks is problematic. This makes them a "special" chemical family in geochemistry, which remains little studied compared to other geochemical tools.

Noble gases in the Earth's parent bodies

Noble gases in meteorites have been studied since the 1960s, yet there are still uncertainties about the isotopic compositions of all the noble gases in the Earth's parent bodies. This is due in particular to the presence in meteorites of cosmogenic isotopes, produced by collisions of high-energy particles with the major atoms of rocks, which significantly modify the early isotopic ratios of noble gases, since these are relatively scarce. The classic link between parent bodies and chondrites (carbonaceous or enstatite) made for many of the chemical elements in the periodic table can possibly also be made for noble gases, but with caution.

As mentioned above, the presence of cosmogenic isotopes significantly masks the primary composition of noble gases in chondrites, making analysis of noble gas composition difficult. This is obvious for neon (Figure 6.2). The Earth's mantle shows high values of the ^{20}Ne/^{22}Ne ratio, close to the solar value. This is not observed in carbonaceous chondrites, which show a lower and rather scattered ^{20}Ne/^{22}Ne ratio, displaying a significant contribution from pre-solar grains ("Neon A") and/or Q phase (see next paragraph) (Huss and Lewis 1995). Ordinary chondrites and enstatite chondrites show neon signatures closer to that of the Earth's mantle

(Javoy et al. 2010; Moreira 2013), but this could reflect the fact that these meteorites are often pieces of ejected regoliths, hence irradiated by the solar wind as is the lunar soil (Eberhardt et al. 1972). The debate about the origin of light gases on Earth and how they were incorporated into Earth's parent bodies will be discussed later in the chapter (section 6.2.2.4).

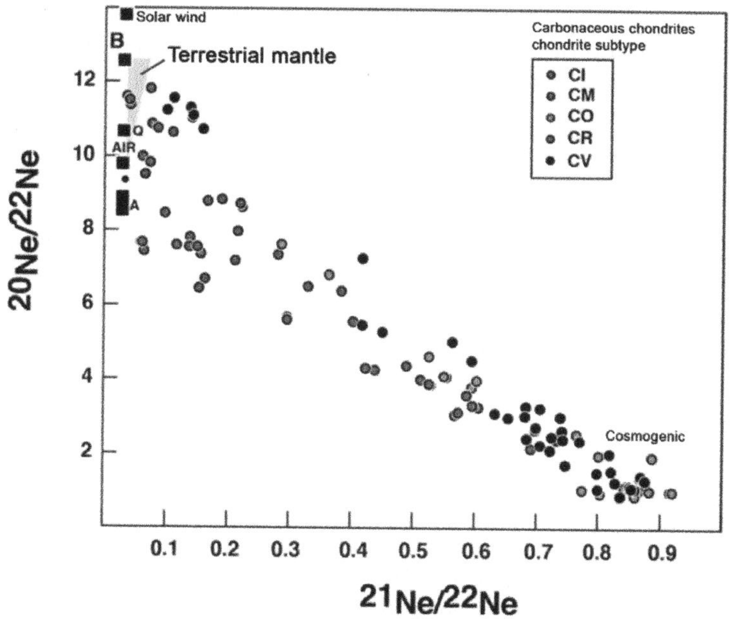

Figure 6.2. *Figure showing the three isotopes of neon in carbonaceous chondrites (figure adapted from Moreira (2013)). For a color version of this figure, see www.iste.co.uk/monteux/earth1.zip*

COMMENT ON FIGURE 6.2.– *Even without the cosmogenic contribution, it is very difficult to establish the neon composition of the Earth's parent bodies from carbonaceous chondrites, given that the value for the Earth's mantle is estimated at between 12.5 and 13.8 (yellow zone in the figure). "Cosmogenic" production reflects interactions between high-energy particles (notably protons) and mineral atoms, producing so-called cosmogenic or spallogenic son isotopes.*

The question seems simpler for heavy noble gases (Ar, Kr and especially Xe), even if work is still in progress to answer the question of their origin on Earth (e.g. Péron et al. 2021). It has been shown since Lewis et al. (1975)

that heavy noble gases are trapped on a component called the "Q phase", which corresponds to the insoluble organic matter of meteorites. This is highly enriched in heavy noble gases, and the heavier the element is, the greater the enrichment, probably reflecting a temperature-dependent adsorption process favoring heavy gases. The isotopic signature of xenon is different in this Q phase and in the solar wind (Wieler et al. 1991; Busemann et al. 2000). This has been explained by a phenomenon of ionization and implantation on this organic matter that fractionates the xenon isotopes in the direction expected (favoring light isotopes) for the xenon isotopic composition of the Q phase (Marrocchi et al. 2011).

Alternatively, it can be proposed that the Q phase simply reflects the isotopic composition of the nebula and that the solar wind, often accepted as reflecting this, is in fact isotopically fractionated for xenon as it is for neon and argon, during the formation of the solar wind itself (Heber et al. 2012). In this case, the heavy noble gases trapped in chondrites reflect a solar composition that will be found in the Earth's interior.

Chondrites contain pre-solar grains, with isotopic signatures reflecting variable stellar nucleosynthesis (Avice et al. 2020). The Kr and Xe isotopic composition of the Earth's parent bodies may therefore vary from that of the Q phase (e.g. Holland et al. 2009) if these heterogeneities are carried by silicates. Earth's composition may also differ from that of chondrites due to heterogeneity in the distribution of pre-solar grains in the inner Solar System (Bouvier and Boyet 2016). It may therefore be irrelevant to simply consider the Q phase as reflecting the composition of the total Earth for heavy noble gases, even though it now seems accepted that it shows a Q phase signature "fragrance" (Holland and Ballentine 2006; Holland et al. 2009; Moreira et al. 2018; Peron and Moreira 2018; Bekaert et al. 2019; Péron et al. 2021).

We can therefore see that there is a difference between the origin of light noble gases (He, Ne) and that of heavy noble gases (Ar-Kr-Xe), perhaps in terms of chronology, but above all in terms of the incorporation process of noble gases in the Earth's interior or in parent bodies. The composition of heavy noble gases in the Earth more or less resembles that of chondrites (and the Q phase), while that of neon (and therefore also probably helium) has an affinity with solar composition. Argon may have an intermediate origin.

The addition of a small proportion of cometary xenon to the Earth's atmosphere has recently been proposed based on measurements of xenon in comet 67P/Churumov-Gerasimenko. Indeed, it has long been observed that the xenon isotopic signature of the Earth's atmosphere requires the addition of a component with a deficit in ^{134}Xe and ^{136}Xe. The addition of a small proportion of cometary xenon to the Earth's atmosphere has recently been proposed based on measurements of xenon in comet 67P/Churumov-Gerasimenko. Indeed, it has long been observed that the xenon isotopic signature of the Earth's atmosphere requires the addition of a component with a deficit in ^{134}Xe and ^{136}Xe relative to the other xenon isotopes in order to explain its isotopic spectrum, whatever the mass fractionation process. This is to explain the present-day composition by fractionation of a theoretical primordial atmosphere called "U-Xe". However, it turns out that xenon analyses on comet 67P/Churumov-Gerasimenko show this deficit in these isotopes from pure "r"-type nucleosynthesis, in contrast to the other xenon isotopes that are produced during "p"-type or even "s"-type nucleosynthetic processes (Marty and Scienceteam 2017). This signature is not observed in chondrites.

Thus, for Marty and Scienceteam (2017), the contribution of ~10% cometary xenon to the Earth's atmosphere is necessary to explain the deficit in the two heavy xenon isotopes of the primordial atmosphere. However, another component, a xenon carrier with a "p"-type nucleosynthetic origin (enriched in light xenon isotopes, but not in heavy ones), must also be added to terrestrial xenon to explain the xenon in Earth's primordial atmosphere; otherwise, the latter would show a significant deficit in ^{124}Xe and ^{126}Xe (Avice et al. 2020).

To date, the carrier of this component has not yet been clearly identified, but it shows that a heterogeneous distribution of grains carrying nucleosynthetic anomalies existed during planet formation. This seems to be the case for other isotopic systems such as Sm or Nd (e.g. Andreasen and Sharma 2006; Burkhardt et al. 2016).

The two light noble gases (He, Ne) have a debated origin on Earth. These are in overabundance relative to concentrations in chondrites, and as discussed in the previous section, the neon isotopic composition of chondrites does not match the composition of the Earth's mantle (Moreira 2013). The latter is of solar "type", i.e. exhibiting an isotopic affinity with the isotopic composition of the solar wind, although its origin on Earth is not yet precisely known (Sarda et al. 1988; Honda et al. 1991; Moreira et al. 1995; Moreira and Allègre 1998).

The two scenarios proposed to explain the origin of helium and neon are as follows:

1) a dissolution of a "solar" atmosphere, captured from the accretion disk during the formation of telluric planets, into an ocean of magma (Sasaki and Nakazawa 1990; Sasaki 1999; Yokochi and Marty 2004; Williams and Mukhopadhyay 2019);

2) an incorporation of solar wind into Earth's parent bodies during the formation of these objects (Trieloff et al. 2000, 2002; Trieloff and Kunz 2005; Raquin and Moreira 2009; Moreira and Charnoz 2016; Jaupart et al. 2017; Peron et al. 2017, 2018; Vogt et al. 2019, 2021). This will be discussed again in section 6.2 under open questions in noble gas geochemistry.

6.2.2. *Noble gas analysis*

6.2.2.1. *Analytical methods and challenges*

Noble gases are analyzed in geological samples using mass spectrometry, whether the mass separator is an electric quadrupole or a magnetic sector. Both are based on the same ionization and ion collection principle; only the mass separators are different (electric for the former, magnetic for the latter). Gases can be extracted from solid samples by melting in an oven by laser extraction (heating or ablation) or vacuum grinding in a mill. The entire gas extraction and purification process takes place under ultra-high vacuum to limit problems of atmospheric contamination. In general, this takes place via a "line", connected directly to the mass spectrometer to avoid moving and storing rare gases for too long before their analysis by mass spectrometry.

In most natural geological samples, ions of all noble gases are detected with an electron multiplier, with the exception of the more abundant ^{4}He and ^{40}Ar, which are measured in a Faraday cage. These ions are produced in an ion source, generally of the Nier type, although other ion sources have been developed in some laboratories (e.g. Baur-Signer at ETH Zurich). They operate on the same general principle of filament ionization, with relatively modest ionization yields given the physical constraints of the ion sources and the large-volume flight tubes.

Measurements of noble gas isotopic ratios are generally less precise than for other chemical elements, mainly due to the quantity of material to be analyzed. To illustrate this, here are a few figures. A gram of ridge basalt, considered "rich" compared to other samples, contains ~10^9 atoms of ^3He, while a gram of olivine from a hotspot basalt contains ~10^6 atoms of ^3He; obviously, these figures are maxima, and the number of atoms to be measured may be much lower. These numbers are even lower for certain isotopes of neon, krypton and xenon. The second reason for reduced precision is the proportion of analytical blank, which can be significant (several %), and the correction it implies, leading to sometimes large uncertainties. Details of analytical techniques specific to noble gas analysis can be found in Zimmerman and Bekaert (2020).

6.2.2.2. *Atmospheric contamination of samples and extraction techniques*

One of the major aspects of noble gas geochemistry is the almost ubiquitous presence of atmospheric gases in samples, despite ultra-high vacuum drying several hours to several days prior to gas extraction and mass spectrometry analysis. In the case of oceanic basalts, several clues suggest that this contamination occurs after eruption and that atmospheric gases are localized in cracks or empty vesicles (Figure 6.3). These are difficult to pump under vacuum (Ballentine and Barfod 2000), perhaps due to the small diameters of the cracks (a few μm) (Roubinet and Moreira 2018). Indeed, bubble-by-bubble analyses by laser ablation, whose internal pressure is controlled during extraction, indicate that mantle-signal vesicles have eruption pressures while atmospheric compositions are associated with low atmospheric pressure, suggesting post-eruption contamination (Raquin et al. 2008; Colin et al. 2015; Peron et al. 2016, 2019).

This surface contamination is problematic when it comes to interpreting the measurements. Indeed, when it comes to the recycling of noble gases in the mantle (Bach and Niedermann 1998; Sarda et al. 1999; Sarda 2004), and therefore to regasification, it is essential to distinguish between surface and deep signals. Avoiding contamination is therefore essential, and there is no magic bullet for this. Stepwise heating (Sarda et al. 2000), stepwise grinding (Moreira et al. 1998) and laser ablation techniques (Burnard et al. 1997; Raquin et al. 2008) have been developed to control atmospheric contamination, but they can never be completely eliminated.

In the case of step-analyzed mantle samples showing solar-type (non-atmospheric) neon isotopic signatures, it is possible to use correlations between $^{20}Ne/^{22}Ne$ and other isotopic ratios. These can then be extrapolated to a fixed $^{20}Ne/^{22}Ne$ value, thus "correcting" for atmospheric contamination (Moreira et al. 1998; Trieloff et al. 2000, 2002; Hopp and Trieloff 2005; Parai et al. 2009, 2012; Mukhopadhyay 2012; Tucker et al. 2012). But this requires knowledge of the $^{20}Ne/^{22}Ne$ value of the mantle (solar, solar wind, etc.) and therefore neglects the possible subduction of atmospheric neon, at least locally in the sample source mantle (Kendrick et al. 2011).

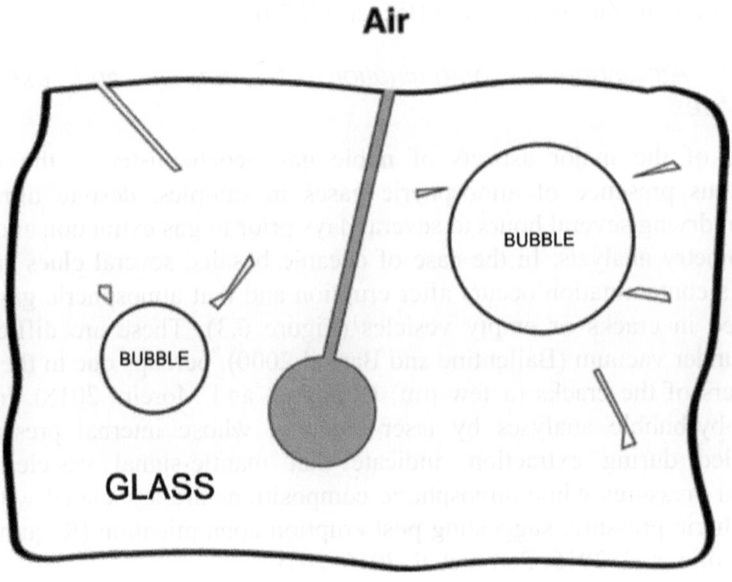

Figure 6.3. *Explanatory diagram of the possible source of ubiquitous atmospheric contamination of magmatic samples. For a color version of this figure, see www.iste.co.uk/monteux/earth1.zip*

COMMENT ON FIGURE 6.3.– *After eruption, some vesicles, connected to the surface, fill with water or air and thus contain "atmospheric" gases. Despite being evacuated and baked at 150°C, these atmospheric gases are difficult or impossible to extract without damaging the integrity of the sample by grinding or heating.*

This atmospheric contamination must therefore always be borne in mind when interpreting measurements of noble gases, including neon. It is a strong limiting factor in understanding the dynamics of exchanges between the surface and mantle.

Current composition of noble gases in the mantle

The main sources of information on the isotopic composition of noble gases in the Earth's mantle are oceanic basalts (hotspots/OIBs and ridge basalts/MORBs) (see Chapter 4 by Cinzia Farnetani). There are others, but they are often more difficult to interpret. Volcanic gases are often a mixture of magmatic gases and meteoric water, which requires corrections and the signal is often difficult to deconvolute, although for helium this remains a possibility. CO_2-rich well gases give fundamental results, but often include a crustal contribution, which again requires correction to determine the primary mantle signal. Mantle enclaves often give relevant results, but the question of their origin is often raised. The same applies to diamonds. In this chapter, we shall confine ourselves to oceanic basalts, whether basaltic glasses or glassy/fluid inclusions in magmatic minerals.

6.2.2.3. *Helium*

Helium is the rare gas that provides the most valuable information on mantle heterogeneity. Several hundred MORB and OIB analyses are available, and a histogram of the $^4He/^3He$ ratio is shown in Figure 6.4. The $^3He/^4He$ ratio can also be found in the literature, generally using the notation R/Ra, where Ra is the atmospheric ratio (= 1.38 10^{-6}). Both notations can be found in the scientific literature. The $^4He/^3He$ notation better represents the radiogenic isotope geochemistry approach, where the ratio between the son and the stable isotope is represented. Historically, researchers studied excess ^3He in seawater, and therefore used the $^3He/^4He$ notation, which has remained in use and is still the most widely used. The histograms in Figure 6.4 show the two ratios in MORB and OIB. There is a clear difference between the two histograms. The MORB source reservoir appears more homogeneous, with an average value around 90,000 (R/Ra ~ 8), and a shoulder toward values of 50,000–60,000 (R/Ra between 10 and 15), which corresponds to MORBs located near hotspots (Iceland, Saint Paul-Amsterdam, Shona-Discovery (Poreda et al. 1986; Moreira et al. 1995; Graham et al. 1999; Sarda et al. 2000)) and therefore under the influence of a source with low $^4He/^3He$.

OIBs show a much more spread-out histogram, with both low $^4He/^3He$ values (around 18,000; R/Ra = 40), notably in Iceland (Graham et al. 1998; Stuart et al. 2003), and high ones (up to 200,000; R/Ra ~ 3.5) observed in some Pacific hotspots, of the HIMU type (Hanyu and Kaneoka 1997), in the Azores, on the island of São Miguel (Kurz et al. 1990; Moreira et al. 1999;

Moreira et al. 2012), and some hotspots on the African rim (Barfod et al. 1999; Class et al. 2005; Day and Hilton 2011).

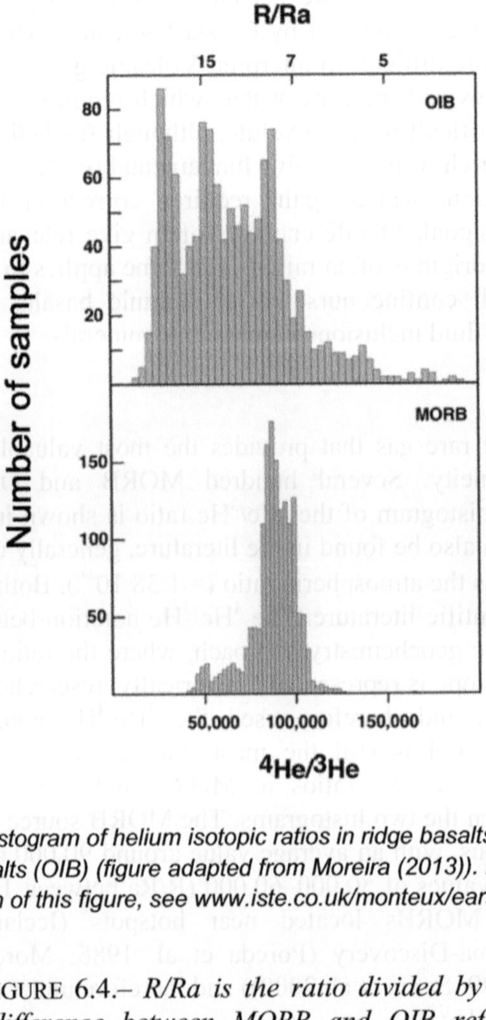

Figure 6.4. *Histogram of helium isotopic ratios in ridge basalts (MORB) and island basalts (OIB) (figure adapted from Moreira (2013)). For a color version of this figure, see www.iste.co.uk/monteux/earth1.zip*

COMMENT ON FIGURE 6.4.– *R/Ra is the ratio divided by the atmospheric ratio Ra. The difference between MORB and OIB reflects a different concentration of 3He in the sources of these basalts (see text).*

The high $^4He/^3He$ values can be understood in the light of other isotopic systems (Sr, Nd, Pb). They certainly reflect interaction with recycled components, continental crust and continental lithospheric mantle. Low values, often referred to as "early ^3He", provide crucial information on the deep mantle.

6.2.2.4. *Neon*

Neon shows much the same systematics as helium, but with greater sensitivity to the primordial signal (Moreira and Allègre 1998). Figures 6.5 and 6.6 show the three neon isotopes in MORBs and OIBs. Figure 6.5 shows that most MORBs are aligned on a straight line ("MORB line"; Sarda et al. 1988). This line represents a mixing line between an atmospheric pole and the mantle pole. This illustrates the atmospheric contamination of oceanic basalts discussed in section 6.2.2.2, which results in a mixing line. The samples above this MORB line are the same ones that show low $^{4}He/^{3}He$ ratios, influenced by "early ^{3}He" hotspots.

The maximum value of the $^{20}Ne/^{22}Ne$ ratio is around 12.5 for the MORB source. This appears to be the upper limit, confirmed by a different methodological approach on CO_2 sink samples (Ballentine et al. 2005). The origin of this 12.5 value will be discussed as it is still a debated scientific issue.

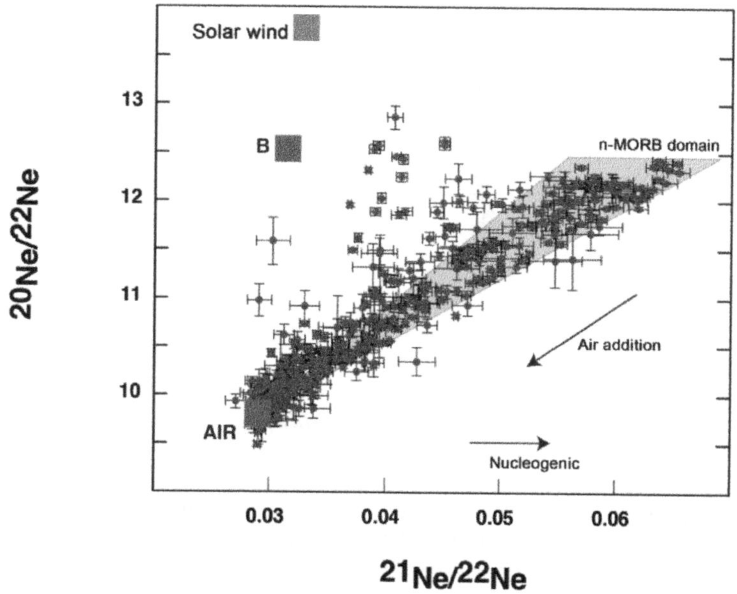

Figure 6.5. *Three-isotope diagram of neon showing a compilation of ridge basalts (MORB) (figure adapted from Moreira (2013)). For a color version of this figure, see www.iste.co.uk/monteux/earth1.zip*

COMMENT ON FIGURE 6.5.– *Most MORBs line up on a mixing line between AIR (or seawater) and a mantle pole with a $^{20}Ne/^{22}Ne$ of ~12.5 and a $^{21}Ne/^{22}Ne$ of ~0.06. This mixing line reflects atmospheric contamination (see section 6.2.2.2). The pole called B represents Neon-B, defined by Black (1971) and representing the implanted solar wind that can be measured in gas-rich meteorites or lunar soil. This value reflects the balance between solar wind implantation and grain sputtering (spatial erosion) (Raquin and Moreira 2009). The solar wind value was determined by analyzing Genesis probe targets exposed to solar wind (Grimberg et al. 2006). This solar wind is isotopically fractionated with respect to the composition of the sun (Heber et al. 2012). Neon values not on the right-hand side of MORBs are MORBs collected on bathymetric anomalies showing OIB-like $^{4}He/^{3}He$ ratios (Moreira et al. 1995; Sarda et al. 2000).*

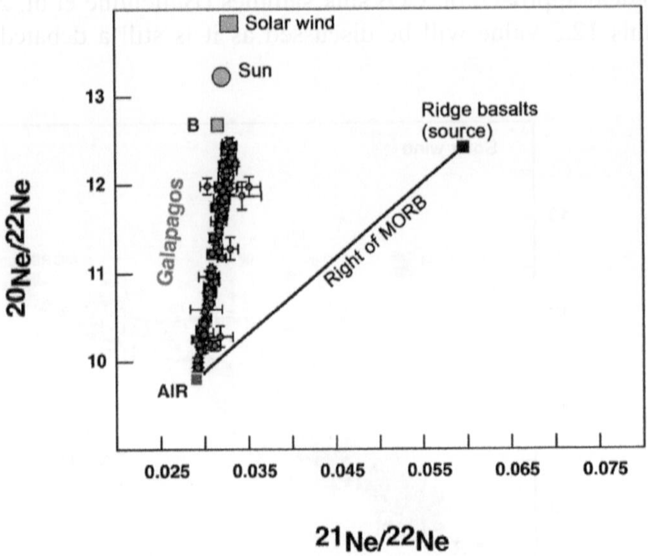

Figure 6.6. *Three-isotope neon diagram showing results from an early helium OIB (Galapagos) (data from Kurz et al. (2009)). For a color version of this figure, see www.iste.co.uk/monteux/earth1.zip*

COMMENT ON FIGURE 6.6.– *What is important to observe in this diagram is the value of the slope of the mixing line. This is greater than the MORB line (Sarda et al. 1988), which means that at an equivalent $^{20}Ne/^{22}Ne$ (Neon B or solar; see Moreira and Charnoz 2016), the $^{21}Ne/^{22}Ne$ is higher in the MORB* *source, as is the $^4He/^3He$ ratio. Most hotspots show slopes similar to that of the Galapagos line, but often intermediate between it and the MORB line.*

Figure 6.6 shows the same neon isotopic ratios, but this time in an early ^3He hotspot – the Galapagos (Kurz et al. 2009). What can be seen in this figure is the value of the slope of the mixing line, which is different from that of the MORBs. The line also reflects the atmospheric contamination of the samples. Because of this contamination, and to avoid it, we need to consider the value of the $^{21}Ne/^{22}Ne$ ratio, for a given value of $^{20}Ne/^{22}Ne$, as that of the mantle. We call this "contamination-corrected" ratio, $^{21}Ne/^{22}Ne_{extrap}$, which can be calculated either by taking a mantle value $^{20}Ne/^{22}Ne = 12.7$ (Moreira and Charnoz 2016), or by taking the solar value (13.8) (Moreira and Allègre 1998). Thus, OIBs and MORBs show a different $^{21}Ne/^{22}Ne_{extrap}$ ratio (~0.035 vs. 0.060, respectively). This difference reflects a higher proportion of nucleogenic ^{21}Ne in the MORB source than in the OIB source.

This will not be discussed here, but it should be noted that the $^{21}Ne/^{22}Ne_{extrap}$ ratio correlates perfectly with the $^4He/^3He$ ratio. However, it shows a hyperbolic, non-linear mixing curve, reflecting a $^3He/^{22}Ne$ ratio that differs in the two sources by a factor of 5–10 (Moreira and Allègre 1998; Moreira et al. 2001; Moreira 2013).

6.2.2.5. *Argon*

Argon has a similar and ultimately quite simple systematics. Argon has three isotopes, two of them stable and non-radiogenic (^{36}Ar, ^{38}Ar) and one radiogenic (^{40}Ar) from the decay of ^{40}K. The $^{38}Ar/^{36}Ar$ ratio is globally homogeneous (to the accuracy of the data) in the Earth's mantle and similar to the value of the atmospheric ratio, and also close to the chondritic value as determined in the "Q phase" (Raquin and Moreira 2009), the carrier of heavy noble gases in chondrites (Busemann et al. 2000). One of the analytical challenges of the next few years will be to detect fine variations in the $^{38}Ar/^{36}Ar$ ratio in the mantle, and to decouple the various possible origins of argon in the Earth's parent bodies (solar, implanted solar wind, Q phase, etc.) and study any possible leakage of argon into space.

The $^{40}Ar/^{36}Ar$ ratio shows very wide variations on Earth. The atmosphere has a ratio of 296, while the least contaminated mantle samples show ratios of over 25,000 for MORBs (Jambon et al. 1985; Sarda et al. 1985; Burnard et al. 1997; Moreira et al. 1998; Trieloff et al. 2003; Raquin et al. 2008; Péron et al. 2019) and a few thousand for OIBs (Valbracht et al. 1997; Harrisson et al. 1999; Trieloff et al. 2000, 2002; Colin et al. 2015).

Neon can be used as a proxy for atmospheric contamination to determine the $^{40}Ar/^{36}Ar$ of MORB and OIB source reservoirs. Moreira et al. (1998) determined a value of ~30,000 for the MORB source mantle, a value confirmed by bubble-by-bubble analysis using laser ablation, a technique that limits atmospheric contamination problems (Raquin et al. 2008; Péron et al. 2019). It should be noted, however, that some authors have proposed that the depleted mantle has a higher $^{40}Ar/^{36}Ar$ ratio, around 40,000 (Marty and Humbert 1997; Tucker et al. 2012), but the uncertainties are high (±9,000), making it difficult to consider this value as representative. Moreover, these values are local estimates since the mantle can be locally heterogeneous if there is recycling of surface material into the mantle (Sarda et al. 1999; Tucker et al. 2012). Systematic work on the global ridge remains to be done to assess the heterogeneity of the $^{40}Ar/^{36}Ar$ ratio in the convective mantle, while freeing ourselves from atmospheric contamination.

For OIBs, a similar approach on the rare OIBs that can be analyzed, that is, those for which glassy, vesicular, and high-depth (pressure) emplaced samples exist, was carried out. In the case of the Galapagos, the $^{40}Ar/^{36}Ar$ ratio of the source has been estimated at 9,000 par (Peron et al. 2016), well below the $^{40}Ar/^{36}Ar$ ratio of the MORB source, but also far from the early value estimated at ~1 (Busemann et al. 2000). The origin of this lower ratio in the MORB source is difficult to interpret, since both the subduction of surface material with an "atmospheric" $^{40}Ar/^{36}Ar$, and the existence of an early mantle that preserved its $^{40}Ar/^{36}Ar$ but mixed with convective mantle may explain this ratio.

6.2.2.6. *Krypton*

Until recently, a section on krypton would have been either non-existent or at best brief. The rare data published did not show any significant difference between the atmosphere, MORBs and OIBs (Trieloff et al. 2000, 2002). Results from CO_2 wells were the only ones to suggest a possible difference between the atmosphere and the mantle, suggesting a chondritic origin for this gas (Holland et al. 2009). The development of gas extraction, purification and separation techniques, coupled with a new generation of mass spectrometers, has confirmed this result on basalt samples from the mantle (Figure 6.7) (Péron et al. 2021). This confirms work on xenon, which shows that these two gases (Kr and Xe) have a chondritic origin, and not a solar origin in the Earth's mantle.

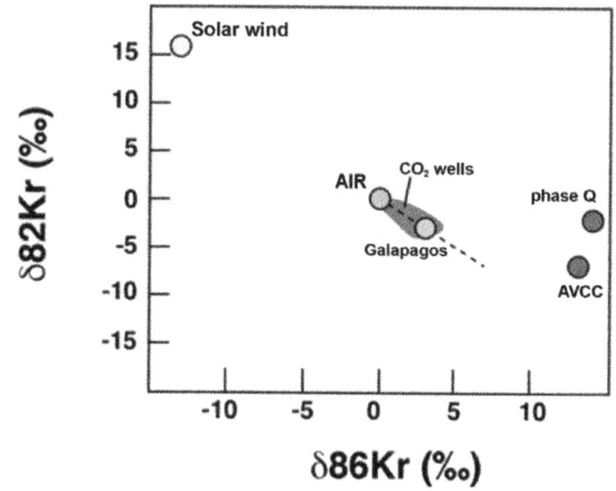

Figure 6.7. *Krypton isotopes in a Galapagos sample showing an excess with respect to air, and suggesting a chondritic origin for krypton (figure adapted from Péron et al. (2021)). For a color version of this figure, see www.iste.co.uk/monteux/earth1.zip*

COMMENT ON FIGURE 6.7.– *AVCC = Average chondric value; phase Q = insoluble organic matter in chondrites (Busemann et al. 2000). CO_2 wells are samples containing gases of mantle origin, enabling precise analyses due to the large quantities of gas available (Holland et al. 2009). This figure is simplified from Péron et al. (2021) for greater clarity.*

6.2.2.7. *Xenon*

Xenon is a particularly interesting noble gas, as it offers a "panoply" of isotopes, both radiogenic and non-radiogenic. This allows us to place strong limits on the evolution of the mantle–atmosphere system.

There are three categories of xenon isotopes:

– *"Stable" isotopes*: ^{124}Xe, ^{126}Xe, ^{128}Xe, ^{130}Xe. These isotopes can therefore be used to trace physical and mixing phenomena without worrying about radiogenic production. These isotopes are scarce and therefore difficult to measure precisely, especially the two lightest ones.

– *Radiogenic isotopes*: ^{129}Xe. It is partly derived from the radioactive decay of ^{129}I, which has a half-life of 17 Myr. It is an extinct radioactivity, and like other short-lived radioactivities, it gives access to the Earth's very early history.

– *Fissiogenic isotopes*: resulting from the spontaneous fission of ^{238}U and ^{244}Pu, another short-lived radioactivity (80 Myr): ^{131}Xe, ^{132}Xe, ^{134}Xe and ^{136}Xe.

Since the pioneering work of J. Reynolds' team, it has been clearly established that the mantle exhibits excesses of radiogenic/fissiogenic isotopes relative to the atmosphere (Butler et al. 1963; Phinney et al. 1978; Smith and Reynolds 1981; Staudacher and Allègre 1982). The fundamental "xenology" diagram is shown in Figure 6.8, which shows the two isotopic ratios $^{136}Xe/^{130}Xe$ and $^{129}Xe/^{130}Xe$ n MORBs and clearly illustrates the excesses of radiogenic isotopes in the mantle.

These excesses are often interpreted as reflecting radiogenic production after mantle degassing.

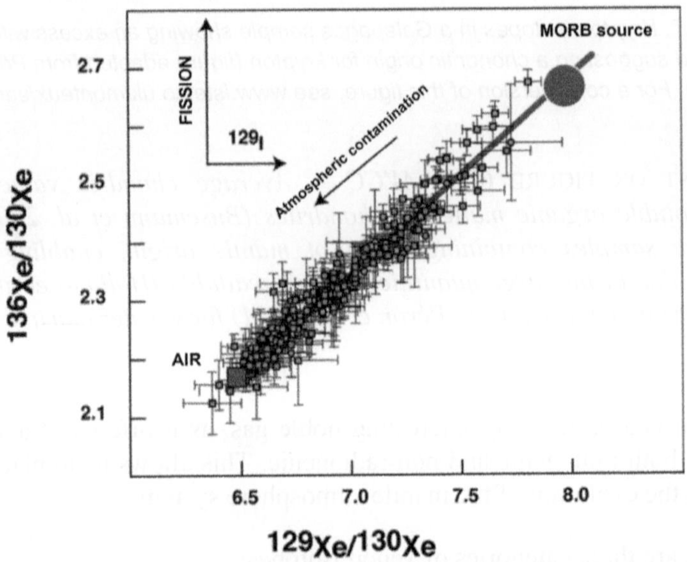

Figure 6.8. *Figure showing for MORBs certain radiogenic xenon isotopes, including ^{129}Xe, the son of ^{129}I which has a half-life of 17 Myr, and ^{136}Xe, resulting from the spontaneous fission of ^{238}U and ^{244}Pu (T½ = 80 Myr) (figure adapted from Moreira (2013)). For a color version of this figure, see www.iste.co.uk/monteux/earth1.zip*

COMMENT ON FIGURE 6.8.– *As with neon, the mixing line reflects atmospheric contamination from a current mantle value. The mantle's ^{129}Xe excess, due to the father's short half-life, is interpreted as very early mantle* *degassing (Staudacher and Allègre 1982). It should be noted that this mantle value around 8 for $^{129}Xe/^{130}Xe$ represents the current value and that this ratio must have been much higher before the onset of subduction (around 20 probably) (Peron and Moreira 2018).*

Xenon isotopic compositions in the OIB source mantle show values close to those of the atmosphere (Trieloff et al. 2000, 2002; Mukhopadhyay 2011). Figure 6.9 shows the $^{129}Xe/^{130}Xe$ ratios calculated using neon as a tracer of atmospheric contamination. The three OIBs shown in this figure (Loihi, Iceland, Galapagos) are "early ^{3}He" hotspots with samples suitable for xenon analysis (fresh glass, great eruption depth/pressure). The origin of these near-air values is difficult to interpret, for the same reasons as for argon. One possible hypothesis is that atmospheric xenon is recycled into the source of these hotspots. Another hypothesis is that these values represent the value of the early mantle, close to the atmospheric value, since this mantle is also rich in xenon.

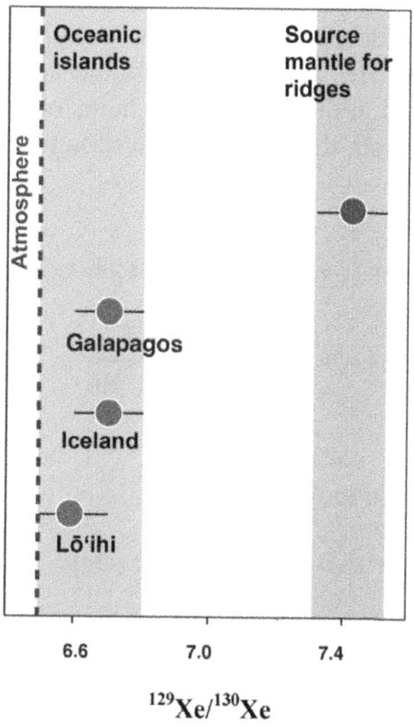

Figure 6.9. *Corrected $^{129}Xe/^{130}Xe$ ratio for atmospheric contamination using neon (see Moreira* et al. *(1998)) in some OIBs and in comparison with the value estimated for the MORB source mantle (from Moreira (2013)). For a color version of this figure, see www.iste.co.uk/monteux/earth1.zip*

The difference between island and ridge basalts

Since Mark Kurz's pioneering work on the isotopic composition of MORBs and OIBs, it has been clear that the source of the islands does not have the same $^{3}He/^{4}He$ ratio as the convective mantle source of the ridge basalts (Kurz et al. 1982a, 1983). This has been confirmed for the radiogenic isotopic ratios of the other noble gases (Ne, Ar and Xe), which consistently show higher ratios in the MORB source (Honda et al. 1991, 1993; Moreira et al. 1995, 1996, 1998; Valbracht et al. 1996, 1997; Moreira and Allègre 1998; Sarda et al. 2000; Trieloff et al. 2000, 2002; Trieloff and Kunz 2005; Mukhopadhyay 2011, 2012; Tucker et al. 2012).

The first notable difference between MORB and OIB is the $^{4}He/^{3}He$ ratio. This is illustrated in Figure 6.4, which shows this isotopic ratio in a series of MORB and OIB samples. We can clearly see that most islands have $^{4}He/^{3}He$ ratios lower than the MORB value (~90,000).

The most commonly accepted hypothesis to explain this difference is that the mantle is rich in ^{3}He, making it possible to "compensate" for the radiogenic ^{4}He produced in the radioactive chains of U and Th, the source of the oceanic islands. Indeed, the equation describing the time evolution of the $^{4}He/^{3}He$ ratio is:

$$\frac{^{4}\mathrm{He}}{^{3}\mathrm{He}} \sim \left(8 \times (e^{\lambda_8 t} - 1) + \frac{7}{137.88} \times (e^{\lambda_7 t} - 1) + 6 \times (e^{\lambda_2 t} - 1)\frac{\mathrm{Th}}{\mathrm{U}}\right) \times \frac{C_U}{^{3}\mathrm{He}}$$

where λ_x are the decay constants of ^{235}U, ^{238}U and ^{232}Th, and C_U is the uranium concentration (in mol/g).

To the first order, for two tanks with the same Th/U ratio (and the same time):

$$\frac{\left(\frac{^{4}\mathrm{He}}{^{3}\mathrm{He}}\right)_{\mathrm{MORB}}}{\left(\frac{^{4}\mathrm{He}}{^{3}\mathrm{He}}\right)_{\mathrm{OIB}}} \sim \frac{C_U^{\mathrm{MORB}}}{C_U^{\mathrm{OIB}}} \times \frac{^{3}\mathrm{He}_{\mathrm{OIB}}}{^{3}\mathrm{He}_{\mathrm{MORB}}}$$

If the isotopic ratio is ~20,000 for OIB and ~90,000 for MORB, then:

$$\frac{^3\mathrm{He}_{\mathrm{OIB}}}{^3\mathrm{He}_{\mathrm{MORB}}} \sim \frac{4.5}{\left(\frac{C_U^{\mathrm{MORB}}}{C_U^{\mathrm{OIB}}}\right)}$$

However, it is commonly considered that the depleted mantle, which is the source of the MORBs, contains at least four times less uranium than the island source. This means that the island source would have to contain at least 18 times more ^{3}He than the MORB source to preserve its low-radiogenic helium signature. This is a simple, first-order calculation, based on the notion of a reservoir depleted/enriched in trace elements (including uranium), which is intended to illustrate why it has always been proposed that the Islands source is enriched in ^{3}He. This says nothing about where this source reservoir of the islands is located. It does, however, seem obvious that if this reservoir is less degassed, it must lie deeper in the Earth's mantle (or even core) than the source of the ridge basalts. It has been located throughout the lower mantle (Allègre and Moreira 2004), at the base of the mantle in the residue of the Basal Magma Ocean (BMO) (Labrosse et al. 2007), possibly even in the core, assuming there is a mechanism for transporting helium from the core to the source of the plumes and that helium is more siderophilic than uranium (Porcelli and Halliday 2001).

The second notable difference between MORB and OIB is neon (Figures 6.5 and 6.6). This can be interpreted in the same way as for helium, insofar as the ^{21}Ne/^{22}Ne isotopic ratio is a "sister" isotopic ratio to that of helium.

Indeed, ^{21}Ne is a so-called "nucleogenic" isotope, resulting mainly from the nuclear reaction ^{18}O(α,n), that is, a fraction of the radiogenic helium produced (5×10^{-8}) reacts with oxygen to produce ^{21}Ne (Moreira and Allègre 1998). This gives us:

$$\begin{aligned}\frac{21\mathrm{Ne}}{22\mathrm{Ne}} &= (21\mathrm{Ne}/22\mathrm{Ne})_0 + \frac{21\mathrm{Ne}*}{22\mathrm{Ne}} \\ &= (21\mathrm{Ne}/22\mathrm{Ne})_0 + \frac{21\mathrm{Ne}*}{4\mathrm{He}*} \times \frac{4\mathrm{He}*}{22\mathrm{Ne}}\end{aligned}$$

So for both reservoirs, knowing that the $^{21}Ne^*/^4He^*$ production ratio is the same in both reservoirs (since it depends only on the major elements), as is the initial isotopic ratio (0.032):

$$\frac{\left(\frac{21Ne}{22Ne}\right)_{MORB} - (21Ne/22Ne)_0}{\left(\frac{21Ne}{22Ne}\right)_{OIB} - (21Ne/22Ne)_0} = \frac{(4He*)_{MORB}}{(4He*)_{OIB}} \times \frac{22Ne_{OIB}}{22Ne_{MORB}}$$

Considering the calculated values of the different reservoirs and the initial value (at a fixed $^{20}Ne/^{22}Ne$ of 12.7, see Peron et al. (2018)), we obtain:

$$^{22}Ne_{OIB} \sim 9{\cdot}3 \times 4 = 37 \times {}^{22}Ne_{MORB}$$

if we always consider a uranium concentration ratio of 4.

The OIB source reservoir therefore contains more primordial neon (a factor of at least 37) than the MORB source. This neon enrichment factor is known to be greater than for helium (Moreira and Allègre 1998). This reflects a different $^3He/^{22}Ne$ ratio in the island source and in the MORB source, although it is not yet known why (Moreira and Allègre 1998). Some authors have proposed that this reflects different stages of magma ocean–atmosphere equilibrium following atmospheric losses (Tucker and Mukhopadhyay 2014).

A similar calculation can be made for the $^{40}Ar/^{36}Ar$ ratio, but the robustness of this calculation lies in neglecting argon subduction, which, as we shall see in the next section, is not an obvious assumption. The $^{40}Ar/^{36}Ar$ ratio of the MORB source is greater than 20,000 (Moreira et al. 1998), whereas it is ~9,000 in that of the islands with an early helium ratio (Peron et al. 2018). If these ratios just reflect $^{40}K/^{36}Ar$ ratios, without evoking recycling, then we also have a higher concentration of primordial argon in the Islands source than in the MORB one. However, caution must be exercised in the case of argon due to possible recycling of atmospheric argon.

Thus, it seems obvious, on the basis of helium and neon isotopic ratios, to postulate the existence of a reservoir richer in these primordial gases – sampled by the mantle plumes that give rise to oceanic islands such as Hawaii, Iceland, the Galapagos Islands and Reunion Island. The origin and physics of the preservation of such a reservoir can be seen in the other chapters of this volume.

6.3. The evolution of the mantle–atmosphere system

There is an abundant literature on mantle degassing using noble gas geochemistry, often coupling argon and xenon, with the addition of helium as a tracer of current degassing fluxes. No exhaustive description will be given of all these articles, which generally use the same constraints, observables and methodological approaches. Instead, the focus here will be on the constraints that exist and the main problems that arise in such studies.

6.3.1. *Outflows: mantle degassing*

To study mantle degassing, it is essential to have constraints on fluxes, in addition to the well-known current isotopic ratios of mantle and atmosphere. The main tool for studying volatile element fluxes out of the Earth's mantle is ^{3}He. The latter is virtually absent in the Earth's atmosphere, as helium escapes thermally into space. Although the detection of ^{3}He fluxes in natural gases goes back a long way (Aldrich and Nier 1946), it is mainly the work on helium dissolved in seawater that has enabled us to estimate the flux of He leaving the mantle (Clarke et al. 1969). From concentrations measured in seawater above the ridge axis, and by establishing circulation models, it is possible to determine a flux of ^{3}He ranging from ~1,000 moles/year (Farley et al. 1995) to 527 ± 102 moles/year (Bianchi et al. 2010).

Another approach, developed by Chavrit et al. (2014), estimates the ^{3}He flux by assuming the average $CO_2/^{3}He$ ratio (~2×10^{9}) and knowing the CO_2 flux. The latter is determined by establishing laws linking vesicularity and geochemical parameters (K/Ti for example) and determining the average vesicularity (= CO_2) of a number of well-characterized ridge segments. The flux estimated by these authors is between 300 and 400 moles/year of ^{3}He, lower than other estimates. This may be linked to the fact that vesicles are surely lost prior to eruption, thus minimizing the CO_2 content determined by Chavrit et al. (2014).

A third approach uses the estimated ^{3}He concentration in the mantle (and therefore in the crust) and the flux of crust produced annually (7×10^{16} g/year; Keller et al. 2017). The mantle has a ^{3}He concentration of the order of 4×10^{-15} mole/g (Moreira and Kurz 2013). Note that some authors propose a concentration twice lower, which obviously changes the calculation below by a factor of 2 (Tucker et al. 2018). Considering that helium is highly incompatible and that the average partial melt rate is 10%, this implies a pre-degassed crustal concentration of 4×10^{-14} mole/g. With a crustal production of 7×10^{16}g/year, this leads to a ^3He flux of 3,125 moles/year. This value is almost certainly overestimated, as the ^3He concentration of the order of 4×10^{-15} mole/g determined by Moreira and Kurz (2013) is predicated on the study by Moreira et al. (1998) based on a non-degassed basalt (popping rock 2πD43). This gives a high concentration for helium compared with all MORBs, and in a somewhat peculiar geological context (Péron et al. 2019). This sample may therefore reflect an accumulation of bubbles in the magma chamber and thus overestimate the helium concentration. This calculation also assumes that the entire crust is massively degassed as it forms, which is indeed suggested by helium measurements in gabbros from oceanic boreholes (Moreira et al. 2003).

As we can see, flux estimations vary by a factor of 10, between 300 and 3,000 moles/year of ^3He leaving the mantle per year, depending on the approach used. Considering the various uncertainties surrounding the assumptions of the different approaches, a value of around 1,000 moles/year for the current ^3He flux seems reasonable.

6.3.1.1. *K-Ar and I-Xe pairs and the two-stage degassing of the mantle*

While the flux of ^3He out of the mantle is a strong constraint, it remains only a current value, and other constraints are needed to study mantle degassing throughout Earth's history. These could be the isotopic ratios of argon and xenon in the mantle AND the atmosphere. The latter is in fact a reservoir that has been accumulating gases degassed from the mantle for 4.5 billion years, and the aim of the models is thus to explain the composition of the atmosphere, in addition to that of the mantle.

Many authors have used the ^{40}K–^{40}Ar and ^{129}I–^{129}Xe systems to study the mantle–atmosphere pair (Allègre et al. 1986b; Turner 1989; Azbel and Tostikhin 1990; Ozima and Zahnle 1993; Pujol et al. 2013). Indeed, these two systems are complementary, as K-Ar is an integrating system given the long half-life of ^{40}K, and the ^{129}I-^{129}Xe system gives access to degassing during the Hadean, as the half-life of ^{129}I is 17 Myr (this isotope is produced in supernovae, and its existence suggests the contribution of supernova grains shortly before the formation of the Solar System). What the results of the various studies show is the need for degassing in two stages, one early in time (4.4–4.5 Gyr) and the other continuous throughout the Earth's history (see Figure 6.10 to illustrate this result). This makes sense in the light of what we know about the Earth's evolution, that is, an accretion phase with a possible magma ocean favoring efficient degassing, followed by a "plate tectonic" phase, less efficient because the volume of molten mantle (and therefore its degassing) involved is low.

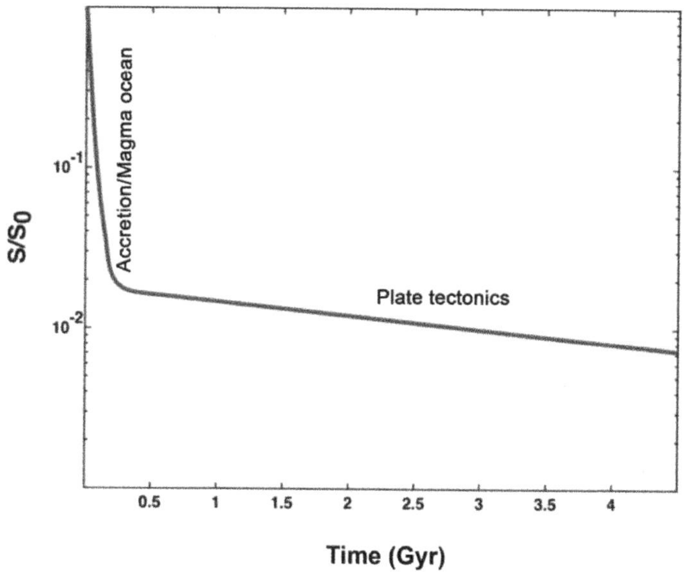

Figure 6.10. *Simple two-stage mantle degassing model (according to Moreira (2013)). For a color version of this figure, see www.iste.co.uk/monteux/earth1.zip*

COMMENT ON FIGURE 6.10.– *The S isotope can be any isotope (^{3}He, ^{130}Xe, ^{36}Ar, etc.). However, this model neglects the subduction of atmospheric gases and potential losses into space.*

6.3.1.2. *Mantle regasification as seen by non-radiogenic xenon isotopes*

One of the difficulties in dealing with mantle degassing is the conjunction of two additional processes to degassing alone. One is the loss of atmosphere to space, and not just helium. Neon could also be affected by a significant loss, thus explaining why the $^{20}Ne/^{22}Ne$ ratio of the atmosphere is not "solar-type" as is that of the mantle (Hunten et al. 1987; Sarda et al. 1988), and xenon as well (Sasaki and Nakazawa 1988; Zahnle et al. 1990).

The other process is the recycling of "atmospheric" gases into the mantle through the deep subduction of materials rich in these atmospheric gases, be it argon (Sarda et al. 1999) or xenon (Holland and Ballentine 2006). Some authors have even proposed that mantle recycling and mantle degassing are steady state (Porcelli and Wasserburg 1995a, 1995b).

As discussed in section 2.3.2, it is very difficult to quantify subduction, since atmospheric contamination of natural samples is always present. Historically, a "subduction barrier" had been proposed for noble gases, particularly xenon. In other words, the inflow of volatile elements into the mantle was negligible, as they had to be extracted to the mantle wedge and then to the surface (Staudacher and Allègre 1988). This is now clearly unfounded, as xenon clearly seems to indicate recycling of atmospheric gases into the mantle. Analyses of krypton isotopic ratios are also beginning to suggest recycling of atmospheric krypton, but the question of the early composition of the Earth remains (Péron et al. 2021). For this reason, we will focus on xenon only.

Xenon has the particularity of having numerous isotopes (nine), some of which are non-radiogenic (and stable). These include ^{124}Xe, ^{126}Xe, ^{128}Xe, ^{130}Xe, and to a lesser extent ^{129}Xe, if we consider that ^{129}I has been extinct for over 4 billion years. These isotopes can therefore be used to trace physical and mixing phenomena without worrying about radiogenic production. However, these isotopes are in short supply. ^{124}Xe and ^{126}Xe are isotopes derived from p-type nucleosynthesis, which produces isotopes of very low abundance. On Earth, these isotopes have abundances of the order of 0.0951% and 0.0887% of xenon, making them very rare and difficult to measure. ^{128}Xe and ^{130}Xe (pure "s"-type nucleosynthesis isotopes) are more abundant, albeit in low proportions (1.9% and 4%, respectively). As a result, measuring the isotopic ratios of these isotopes in mantle samples was for a long time unusable because it was too imprecise. Only analyses of gas samples from CO_2 wells seemed to suggest a composition of the mantle different from that of the atmosphere, which may seem surprising given that these isotopes are non-radiogenic (Cafee et al. 1999; Holland and Ballentine 2006). Volcanic gases can also sometimes indicate something similar, but they are often contaminated by meteoric water, making the excesses weaker (Moreira et al. 2018; Bekaert et al. 2019).

This difference between mantle and atmosphere was recently confirmed on a MORB by Peron and Moreira (2018) using a new gas accumulation analytical protocol that delivers analyses that are less imprecise than before (Figure 6.11). This result confirms the subduction of atmospheric xenon into the mantle, from an early chondritic composition. Moreover, the model proposed by Peron and Moreira (2018) suggests that the onset of subduction of this atmospheric xenon could only have begun 3 billion years ago. The proportion of atmospheric xenon in the mantle is around 90%. The current composition of the Earth's atmosphere is not chondritic due to a loss of xenon to space with isotopic fractionation (Pujol et al. 2011; Avice et al. 2017; Avice et al. 2018), the physics of which will be discussed later in section 5.5 on open questions in noble gas geochemistry. This loss appears to be continuous over geological time, producing an atmospheric composition that is highly isotopically fractionated compared with the primordial composition (chondritic or "U-Xe"). This mantle–atmosphere difference helps limit atmospheric xenon subduction.

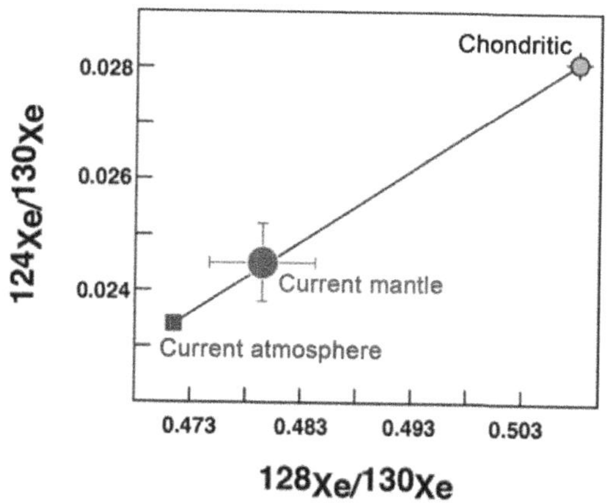

Figure 6.11. *Non-radiogenic xenon isotopes in a gas-rich MORB, with a new analytical protocol to accumulate sufficient xenon for relatively accurate analysis (figure adapted from Peron and Moreira 2018). For a color version of this figure, see www.iste.co.uk/monteux/earth1.zip*

COMMENT ON FIGURE 6.11.– *This sample thus gives the xenon isotopic composition of the present-day mantle, which differs from that of the atmosphere and the chondritic pole. This suggests massive recycling of atmospheric xenon into the mantle (90%), confirming results obtained on CO_2 sinks (Cafee et al. 1999; Holland and Ballentine 2006) and, to a lesser* *extent, data obtained on volcanic gases (not shown for the sake of clarity) (Moreira et al. 2018; Bekaert et al. 2019).*

Possible carriers of atmospheric xenon are serpentinite (Krantz et al. 2019), hydrated minerals and weathered oceanic crust. These appear to be enriched in heavy noble gases (Kendrick et al. 2011; Chavrit et al. 2016), although the effect of the subduction zone filter on the volume of volatile elements in the slab is poorly understood.

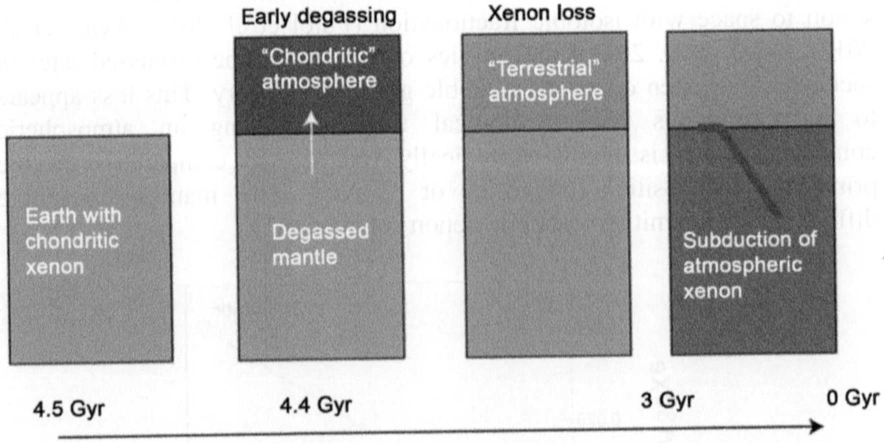

Figure 6.12. *Explanatory diagram of the aging of the xenon isotopic composition in the mantle--atmosphere system. For a color version of this figure, see www.iste.co.uk/monteux/earth1.zip*

COMMENT ON FIGURE 6.12.– *The transition from "chondritic" to "terrestrial" atmosphere appears to have proceeded continuously over geological time, with a possible halt at the time of atmospheric oxygenation (Pujol et al. 2011; Avice et al. 2017, 2018). This loss of xenon, accompanied by isotopic fractionation, is still poorly understood (Zahnle et al. 2019).*

The scenario describing the evolution of the composition of mantle xenon is shown in Figure 6.12. The early Earth must have had a chondritic isotopic signature for xenon. The early atmosphere must also have had a chondritic isotopic signature (referring here to non-radiogenic isotopes). For more than 2 billion years, xenon escaped from the atmosphere (Avice et al. 2018), producing a xenon deficit in the atmosphere (missing xenon). It also caused significant isotopic fractionation, depleting xenon in its light isotopes. It is also possible that 10% of the atmospheric xenon was contributed by comets, which helps explain the slight deficit in ^{134}Xe and ^{136}Xe isotopes in the early atmosphere (Marty and Scienceteam 2017). Efficient subduction of atmospheric xenon would have begun 3 billion years ago, helping to produce the current composition of the mantle (Peron and Moreira 2018).

6.4. Past and present dynamics of the Earth's mantle

6.4.1. *A homogeneous asthenosphere through convection and/or efficient melting?*

The isotopic homogeneity of MORBs, compared with OIBs, has been known since the 1980s (Allègre et al. 1986a; Allègre 1987; Hart 1988; Hofmann 1997), and there are two reasons for this. The first is that the source of OIBs contains recycled material of varying age that has experienced the filter of subduction zones (terrigenous sediments, pelagic sediments, weathered oceanic crust, etc.), and is therefore by nature necessarily highly heterogeneous (Hart and Zindler 1989; Chauvel et al. 1992; Elliott et al. 1999; Eisele et al. 2002; Jackson and Dasgupta 2008). OIBs are also more heterogeneous because the mantle plume also takes along convective mantle as it forms and rises (Hart et al. 1992; Hauri et al. 1994).

The explanation for the greater homogeneity of MORBs compared with OIBs is that the asthenosphere is mechanically mixed by convection (Allègre et al. 1984; Kellogg and Turcotte 1986; Hart 1988; Allègre and Lewin 1995). A possible tracer for understanding and quantifying this mixing is the value of the standard deviation of isotopic ratios, as proposed by Claude Allègre and his team (Allègre et al. 1995; Allègre and Lewin 1995).

Figure 6.13 shows the standard deviation of the ^{4}He/^{3}He ratio in relation to the inverse of the plate spreading rate, which could be an indicator of the strength of convection (Allègre et al. 1995). However, it should be noted that in these works, the authors do not assume that melting under the ridge can also have a homogenizing effect. Indeed, it can be assumed that a high partial melt rate can also result in a high plate spreading rate. These two phenomena combined can thus produce effective homogenization of MORBs.

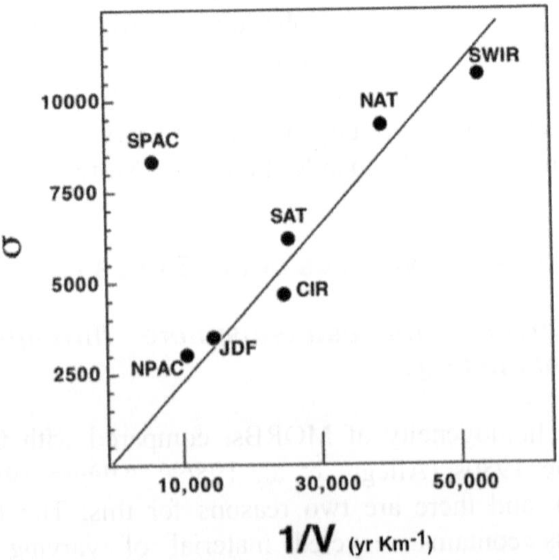

Figure 6.13. *Figure showing the standard deviation of the $^4He/^3He$ ratio of several ridge segments in relation to the inverse of the plate spreading rate (figure adapted from Allègre et al. (1995))*

COMMENT ON FIGURE 6.13.– *The authors interpret this relationship as a reflection of the intensity of mixing in the mantle. A "fast" mantle mixes well, resulting in a low standard deviation of the helium isotopic ratio.*

The example of the Pacific-Antarctic Ridge is symbolic of the fact that mixing by convection is not the only homogenization criterion. Indeed, if we study the 42°S–52°S segment along the Pacific-Antarctic Ridge, we notice that along the axis, the $^4He/^3He$ isotopic ratio is very homogeneous, and it is in fact one of the segments of the global ridge with the smallest standard deviation of this ratio (Moreira et al. 2008) (Figure 6.14).

What is noteworthy, however, is the existence of off-axis volcanism to the north and south of the Ménard transform fracture, which is the result of a difference in spreading rates between the north and south of this fracture zone (Briais et al. 2009).

This volcanism is very recent, and some of the volcanic edifices are even of zero age. Figure 6.14 shows the $^4He/^3He$ ratio of this off-axis volcanism. The $^4He/^3He$ ratios are more radiogenic for some of the off-axis volcanic formations (this is also the case for the $^{206}Pb/^{204}Pb$ ratio) than the mean on-axis value (Moreira et al. 2022). The source of this volcanism is the same mantle as that below the axis, but with a lower partial melt rate than at the axis due to the thickness of the off-axis lithosphere, which limits mantle ascent (see in Figure 6.14). This clearly shows that melting at a high melt rate of a marbled cake-type mantle (Allègre and Turcotte 1986) erases heterogeneities at the axis. However, these remain visible when the partial melt rate is lower, favoring the melting of pyroxenite (corresponding to recycled material) over peridotite (Hirschmann and Stolper 1996).

Therefore, while mixing by convection mechanically homogenizes the mantle, melting at the axis seems to increase this homogeneity.

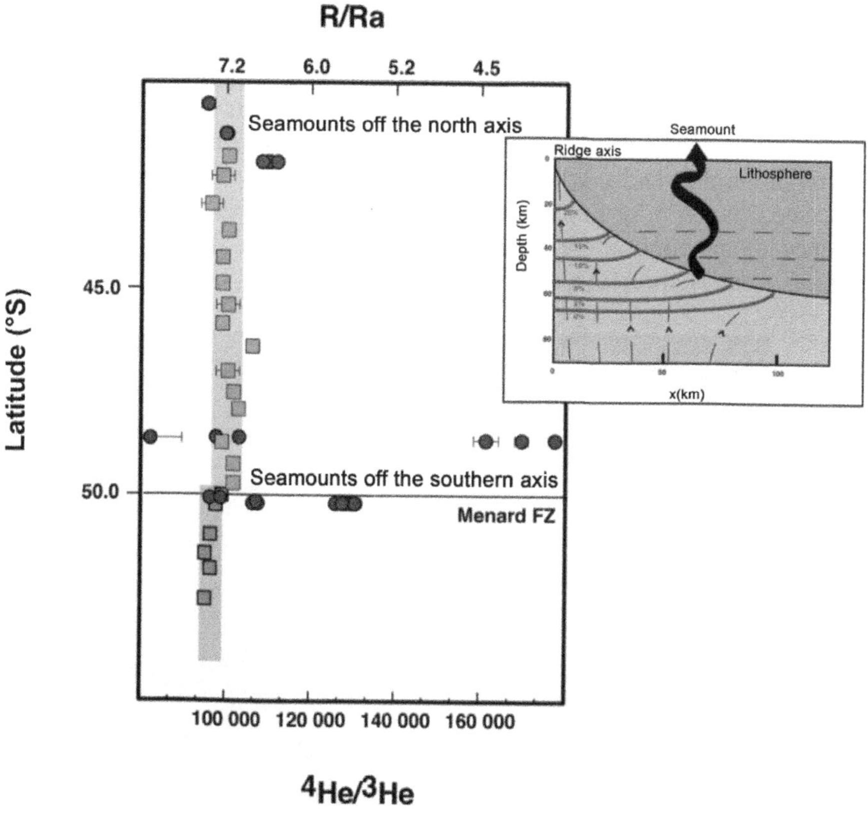

Figure 6.14. *$^4He/^3He$ ratio as a function of latitude for off-axis samples of the Pacific-Antarctic Ridge compared with mean values at the axis (Moreira* et al. *2008). For a color version of this figure, see www.iste.co.uk/monteux/earth1.zip*

COMMENT ON FIGURE 6.14.– *The ratios are radiogenic for some off-axis samples (as are the* $^{206}Pb/^{204}Pb$ *ratios), which can be interpreted as reflecting preferential melting of recycled material, as off-axis volcanism is limited by the thickness of the lithosphere (see the figure and text).*

6.4.2. Noble gas constraints on magma ocean dynamics

The systematic study of noble gases in mantle samples and their comparison with the atmosphere make it possible to discuss the origin of volatile elements on Earth and to study the major geochemical cycles, in particular the degassing and regassing of the mantle. Some of the noble gas isotopes, including ^{129}Xe, son of ^{129}I (T½ =17 Myr), also enable us to address the question of the Hadean Earth, and in particular the early phase of the magma ocean. This magma ocean (Figure 6.15), which resulted from terrestrial accretion and certainly from the giant impacts that occurred at the end of accretion, had a peculiar dynamic characterized by a very high Rayleigh number and vigorous convection (Solomatov 2007; Samuel 2012) (also see Chapter 3 by Julien Monteux and Denis Andrault). The presence or absence of an atmosphere also has repercussions on the duration of this magma ocean (Lebrun et al. 2013).

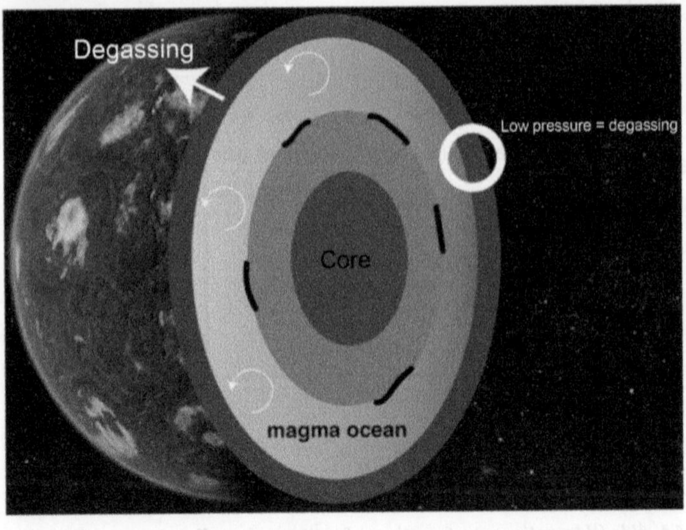

Figure 6.15. *Drawing of the early Earth's magma ocean and the corresponding degassing. For a color version of this figure, see www.iste.co.uk/monteux/earth1.zip*

COMMENT ON FIGURE 6.15.– *Volatile elements are degassed only at low pressure (<100 km). However, convection is very efficient, allowing significant degassing of the silicate Earth (Lebrun et al. 2013) (see Chapter 3).*

Considering the dynamics of this magma ocean, we could suggest that massive degassing may have taken place – although the life span of a magma ocean and its crystallization by the environment may have partially limited complete mantle degassing, thus preserving a primitive reservoir of noble gases in the deep mantle (Labrosse et al. 2007). Furthermore, it should be pointed out that a recent study (Salvador and Samuel 2022) shows that magma ocean degassing was quite low.

We saw earlier that the mantle degassed in two stages, the first of which corresponds to this early phase in the Earth's history. Let us now return to the first phase.

To illustrate the process in a simple way, consider that degassing has mainly taken place in a single stage that follows a law of the type dS/dt = –Ψ S, where S is a stable isotope and Y is a constant degassing term (in year^{-1}), which covertly encompasses the physics of degassing, including solid/magma partition coefficients. For the son, we also assume dF/dt = –ΨF + λP with the same Ψ since this is a term that depends only on physical parameters and the isotopes have the same partition coefficient. P is the father and λ the decay constant. It can be shown that the isotopic ratio R of the mantle is:

$$R(t) = R_0 - \frac{\lambda}{(\psi - \lambda)} \frac{P_0}{S_0} (1 - \exp((\psi - \lambda)t)$$

The atmospheric isotopic ratio:

$$Ra(t) = R_0 + \frac{1}{(\psi - \lambda)} \frac{P_0}{S_0} \frac{(\psi(1 - \exp(-\lambda t)) - \lambda(1 - \exp(-\psi t)))}{1 - \exp(-\psi t)}$$

Therefore, for a very long time:

$$Ra \sim R_0 + \frac{P_0}{S_0}$$

In the case of xenon, S is ^{130}Xe, F is ^{129}Xe, P_0 is the initial ^{129}I, given that $(^{129}I/^{127}I)_0$ is ~10^{-4} for the initial Solar System (Podosek 1970). However, it should be noted that the value that should apply in the case of the Earth is unknown because its formation chronology is imprecise. Nevertheless, the initial P_0/S_0 ratio can be determined in this simple model as R_a-R_0 = 0.5. The unknown in this model is thus the value of Ψ, which reflects the intensity of degassing – a factor of interest here in constraining the dynamics of the early Earth.

The $^{129}Xe/^{130}Xe$ ratio in the mantle prior to atmospheric xenon subduction is estimated to be around ~20 (Chapter 7 of Moreira (2013)). The Ψ value that helps explain this mantle value is ~3.9×10^{-8} years^{-1}. This early degassing must therefore have lasted around 100 Myr (considering that the stable isotope is 99% degassed). The lifetime of a magma ocean is relatively short, around a few million years at most if a dense atmosphere existed and "slowed down" cooling and therefore crystallization (Lebrun et al. 2013). The 100 Myr result therefore indicates, as indeed also mentioned by Lebrun et al. (2013), that the magma ocean and therefore degassing are maintained by continuous accretion or by several successive giant impacts. In fact, repeated giant impacts can explain the $^3He/^{22}Ne$ ratio of the upper mantle, which is different from that of the early mantle (Tucker and Mukhopadhyay 2014).

This is a first-order calculation that overlooks losses to space during impacts, or possible transfer of the father element to surface reservoirs (crust, or even atmosphere rich in moderately volatile elements (Dhaliwal et al. 2018), or loss to space). There are more detailed models that the reader can take a look at to delve deeper into the subject (e.g. Avice and Marty 2014).

6.5. Open questions on the origin and evolution of terrestrial noble gases

There are still several open questions concerning the origin and evolution of noble gases on Earth, as well as their distribution in terrestrial reservoirs and their link with mantle dynamics. The three main questions are listed below and involve three of the noble gases: helium, neon and xenon.

6.5.1. *Helium concentration in island basalts*

One question that has yet to be satisfactorily answered in noble gas geochemistry is the amount of helium measured in oceanic basalts. This is low compared with MORBs (Breddam et al. 2000; Moreira et al. 2001; Kurz et al. 2009) and, above all, with what we would expect from an "early" noble gas reservoir, that is, a factor of ~20 compared with the depleted mantle.

This paradox, brought to light by Anderson (Anderson 1993, 1998), is still not understood. It has often been proposed that it is a difference in degassing that would explain the difference in concentration between MORB and OIB, with a preferential loss of helium in OIB magmas (Gonnermann and Mukhopadhyay 2007; Hopp and Trieloff 2008). However, this is not convincing given the $^4He/^{40}Ar$ ratios of the samples, this ratio being a tracer of the degassing process (Moreira and Sarda 2000; Moreira and Kurz 2013). The helium diffusion that can occur during partial mantle melting could also explain this difference, or at least call into question our knowledge of mantle concentrations (Burnard 2004).

The authors who raised this paradox challenged the notion of a "^3He-rich" mantle in the island source. They proposed that the island source revealed a reservoir with a low $U/^3He$ ratio, as it was very depleted in U relative to ^3He. This could be the lithosphere mantle that is the source of hotspot magmatism (Anderson 1998) or recycled lithosphere mantle (Coltice and Ricard 1999). This would imply that uranium is much more incompatible than helium during partial melting, leaving a residue with low $U/^3He$, "freezing" the helium isotopic ratio at the time of this phenomenon. If this is so, then the $^3He/^4He$ isotopic ratio of this residual mantle must be higher than that of the MORB source mantle, which is what is observed; it must also have a lower helium concentration, as helium is still incompatible.

The main argument against these scenarios is that the trace element signature of the island source is "enriched", not depleted as must be the residual mantle from partial melting. On the other hand, it is not yet clear whether helium is indeed less incompatible than uranium, given the difficulty of determining the helium partition coefficient during partial melting (Marty and Lussiez 1993; Brooker et al. 2003; Parman et al. 2005; Heber et al. 2007).

Thus, the question of the ^{3}He content of the island source mantle is one of the major issues in noble gas geochemistry that needs to be addressed specifically in the light of advances in oceanic island source geophysics.

6.5.2. *The origin of terrestrial neon (and of helium and hydrogen)*

The question of the origin of volatile elements on Earth is a recurring one in the Earth sciences because of the technical advances that are changing the way we see them. This is particularly true of neon, where the precision of isotopic ratios has improved significantly over the last 15 years. Knowing the origin of neon is far from trivial, as it has major cosmo-chemical implications.

Since the 1980s, it has been known that neon is of solar origin in the Earth's mantle (Sarda et al. 1988; Honda et al. 1991). However, this "solar origin" may reflect several scenarios with significant cosmo-chemical consequences. Very early on, this solar origin was interpreted as the presence of a captured solar atmosphere around the early Earth, similar to that of giant planets (Sasaki and Nakazawa 1990; Sasaki 1999; Yokochi and Marty 2004). This has recently been supported by precise analyses of neon in hotspot-influenced oceanic basalts, which showed a possible solar nebula-like value in the mantle (Williams and Mukhopadhyay 2019). On the contrary, the apparent upper limit of the ^{20}Ne/^{22}Ne ratio at ~12.7 in the vast majority of samples and close to the value observed in gas-rich meteorites and lunar regolith may suggest that it was a solar wind implantation phenomenon that brought the light rare gases to Earth (Trieloff et al. 2000, 2002; Raquin and Moreira 2009; Moreira and Charnoz 2016; Peron et al. 2017; Vogt et al. 2019, 2021). Both processes feature solar-like neon (and helium) input, but with two different physical mechanisms.

Solar gas capture is a problematic scenario in terms of chronology, as the gas in the accretion disc disappears in a few million years (Hillenbrand 2005). A planet the size of Earth only forms in a few tens of millions of years, making it impossible to capture gas in a massive way that is significant for the abundance of neon on Earth (Jaupart et al. 2017). Should this scenario prove to be the correct one, it would call into question the chronology of telluric planet formation. Moreover, the atmospheres captured must withstand large impacts, which are not yet clearly established (Genda and Abe 2003; Biersteker and Schlichting 2019). Finally, it has not yet been proven that neon does not fractionate isotopically during vesiculation as magmas rise, which could explain some $^{20}Ne/^{22}Ne$ ratio values above 12.7.

In the case of the solar wind implantation in dust scenario (to allow the incorporation of a large amount of solar neon), the question of how this neon can survive the formation of planetesimals and planetary embryos that undergo thermal metamorphism, or even fusion, due to the presence of ^{26}Al in these objects, is unclear. It should be noted, however, that a recent paper presenting hydrogen data in samples from the asteroid Itokawa suggests that a significant fraction of terrestrial water is derived from this solar wind implantation phenomenon (Daly et al. 2021). However, in this case, the authors propose a later glaze with such material, whereas solar-type neon is present in the source of the islands, and thus appears to be concurrent with accretion. So, irradiation by the solar wind would have taken place before the formation of the parent bodies. The cosmo-chemical implications of this scenario relate to irradiation conditions in the early Solar System, which may favor a chemical sorting of dust, as irradiation modifies the chemical composition of grains, and thus that of Earth's future parent bodies (Bradley 1994; Carrez et al. 2002).

6.5.3. *Missing xenon*

The Earth's atmosphere shows a deficit of xenon when compared with noble gases and chondrites. This "missing xenon" has been the subject of numerous studies, since it seems difficult to explain how the heaviest gas could have been lost without affecting the other gases (Bernatowicz et al. 1984; Sanloup et al. 2005). It should also be pointed out that this missing xenon is associated with significant isotopic fractionation of atmospheric xenon.

This dual phenomenon is still poorly understood. However, some clues seem to be emerging since the discovery that this isotopic fractionation seems to have been taking place throughout geological time, or at least over a large part of the Earth's history. Indeed, it was briefly discussed in section 6.3.1.2 that the xenon isotopic composition of the Earth's atmosphere mainly reflected isotopic fractionation from an early ("U-Xe") composition (Reynolds 1960; Pepin 1991), which was itself a mixture of several compositions (chondritic, cometary, presolar grains) (Pepin 1991; Marty and Scienceteam 2017; Avice et al. 2020). This fractionation can only be explained by a massive loss of xenon, with a preferential loss of light xenon isotopes. The problem is that this only seems to affect xenon and not the other lighter gases. It also seems to affect the atmosphere of Mars in the same proportions (Mathew and Marti 2001), which is also problematic in view of the significant differences in mass between the two planets.

The major observation about atmospheric xenon is that its isotopic composition appears to have varied on Earth since 4.5 Gyr. Indeed, measurement of the isotopic composition of xenon in Archean quartz and barite seems to show an evolution from a chondritic (or U-Xe) composition to a present-day composition (Pujol et al. 2011; Avice et al. 2017, 2018). This has been explained by a hydrodynamic escape of hydrogen into an atmosphere with specific chemical compositions and in an environment of strong UV irradiation, implying that the Earth has lost much of its water (Zahnle et al. 2019). However, at this stage, this model remains preliminary and, above all, does not take into account the observation that Earth and Mars have the same xenon isotopic composition in their atmospheres.

So, having an answer to the question of the "missing" xenon is crucial to addressing the evolution of mantle degassing (and regassing) and providing strong constraints on mantle dynamics, and more specifically on the onset of deep subduction of volatile elements.

6.6. References

Aldrich, L.T. and Nier, A.O. (1946). The abundance of 3He in atmospheric and well helium. *Physical Review*, 70, 983–986.

Allègre, C.J. (1987). Isotope geodynamics. *Earth and Planetary Science Letters*, 86, 175–203.

Allègre, C.J. and Lewin, E. (1995). Isotopic systems and stirring times of the Earth's mantle. *Earth and Planetary Science Letters*, 136, 629–646.

Allègre, C.J. and Moreira, M. (2004). Rare gas systematics and the origin of oceanic islands: The key role of entrainment at the 670 km boundary layer. *Earth and Planetary Science Letters*, 228, 85–92.

Allègre, C.J. and Turcotte, D. (1986). Implications of a two-component marble-cake mantle. *Nature*, 323, 123–127.

Allègre, C.J., Hamelin, B., Dupré, B. (1984). Statistical analysis of isotopic ratios in MORB: The mantle blob cluster model and the convective regime of the mantle. *Earth and Planetary Science Letters*, 71, 71–84.

Allègre, C.J., Hamelin, B., Provost, A., Dupré, B. (1986a). Topology in isotopic multispace and origin of mantle chemical heterogeneities. *Earth and Planetary Science Letters*, 81, 319–337.

Allègre, C.J., Staudacher, T., Sarda, P. (1986b). Rare gas systematics, formation of the atmosphere, evolution and structure of the Earth's mantle. *Earth and Planetary Science Letters*, 81, 127–150.

Allègre, C.J., Moreira, M., Staudacher, T. (1995). 4He/3He dispersion and mantle convection. *Geophysical Research Letters*, 22, 2325–2328.

Anderson, D.L. (1993). Helium-3 from the mantle: Primordial signal or cosmic dust? *Science*, 261, 170–176.

Anderson, D.L. (1998). A model to explain the various paradoxes associated with mantle noble gas geochemistry. *Proc. Natl. Acad. Sci. USA*, 95, 9087–9092.

Andreasen, R. and Sharma, M. (2006). Solar nebula heterogeneity in p-process samarium and neodymium isotopes. *Science*, 314, 806–809.

Avice, G. and Marty, B. (2014). The iodine–plutonium–xenon age of the Moon–Earth system revisited. *Phil. Trans. R. Soc A*, 372, 20130260.

Avice, G., Marty, B., Burgess, R. (2017). The origin and degassing history of the Earth's atmosphere revealed by Archean xenon. *Nature Communications*, 8, 15455.

Avice, G., Marty, B., Burgess, R., Hofmann, A., Philippot, P., Zahnle, K., Zakharov, D. (2018). Evolution of atmospheric xenon and other noble gases inferred from Archean to Paleoproterozoic rocks. *Geochimica et Cosmochimica Acta*, 232, 82–100.

Avice, G., Moreira, M., Gilmour, J.D. (2020). Xenon isotopes identify large-scale nucleosynthetic heterogeneities across the solar system. *The Astrophysical Journal*, 889, 68.

Azbel, I.Y. and Tostikhin, I.N. (1990). Geodynamics, magmatism, and degassing of the Earth. *Geochimica et Cosmochimica Acta*, 54, 139–154.

Bach, W. and Niedermann, S. (1998). Atmospheric noble gases in volcanic glasses from the southern Lau basin: Origin from the subducting slab? *Earth and Planetary Science Letters*, 160, 298–309.

Ballentine, C.J. and Barfod, D. (2000). The origin of air-like noble gases in MORB and OIB. *Earth and Planetary Science Letters*, 180, 39–48.

Ballentine, C.J., Marty, B., Lollar, B.S., Cassidy, M. (2005). Neon isotopes constrain convection and volatile origin in the Earth's mantle. *Nature*, 433, 33–38.

Barfod, D.N., Ballentine, C.J., Halliday, A.N., Fitton, J.G. (1999). Noble gases in the Cameroon line and He, Ne and Ar isotopic composition of high μ (HIMU) mantle. *Journal of Geophysical Research*, 104, 29509–29527.

Bekaert, D., Broadley, M.W., Caracausi, A., Marty, B. (2019) Novel insights into the degassing history of Earth's mantle from high precision noble gas analysis of magmatic gas. *Earth and Planetary Science Letters*, 525, 115766.

Bernatowicz, T.J., Podozek, F.A., Honda, M., Kramer, F.E. (1984). The atmospheric inventory of xenon and noble gases in shales: The plastic bag experiment. *Journal of Geophysical Research*, 89, 4597–4611.

Bianchi, D., Sarmiento, J.L., Gnanadesikan, A., Key, R.M., Schlosser, P., Newton, R. (2010). Low helium flux from the mantle inferred from simulations of oceanic helium isotope data. *Earth and Planetary Science Letters*, 297, 379–386.

Biersteker, J.B. and Schlichting, H.E. (2019). Atmospheric mass-loss due to giant impacts: The importance of the thermal component for hydrogen–helium envelopes. *Monthly Notices of the Royal Astronomical Society*, 485, 4454–4463.

Black, D.C. (1971). Trapped neon-argon isotopic correlations in gas rich meteorites and carbonaceous chondrites. *Geochimica et Cosmochimica Acta*, 35, 230–235.

Bouvier, A. and Boyet, M. (2016). Primitive solar system materials and Earth share a common initial 142Nd abundance. *Nature*, 537, 399–402.

Bradley, J. (1994). Chemically anomalous, preaccretionally irradiated grains in inter-planetary dust from comets. *Science*, 265, 925–929.

Breddam, K., Kurz, M.D., Storey, M. (2000). Mapping out the conduit of the Iceland mantle plume with helium isotopes. *Earth and Planetary Science Letters*, 176, 45–55.

Briais, A., Ondréas, H., Klingelhoefer, F., Dosso, L., Hamelin, C., Guillou, H. (2009). Origin of volcanism on the flanks of the Pacific-Antarctic ridge between 41°30'S and 52°S. *Geochem. Geophys. Geosyst.*, 10, Q09001.

Brooker, R.A., Zu, Z., Blundy, J.D., Kelley, S.P., Allan, N.L., Wood, B.J., Chamorro, E.M., Wartho, J.A., Purton, J.A. (2003). The 'zero charge' partitioning behaviour of noble gases during mantle melting. *Nature*, 423, 738–741.

Burkhardt, C., Borg, L.E., Brennecka, G.A., Shollenberger, Q.R., Dauphas, N. (2016). A nucleosynthetic origin for the Earth's anomalous (super 142) Nd composition. *Nature*, 537, 394–398.

Burnard, P. (2004). Diffusive fractionation of noble gases and helium isotopes during mantle melting. *Earth and Planetary Science Letters*, 220, 287–295.

Burnard, P., Graham, D., Turner, G. (1997). Vesicle-specific noble gas analyses of "popping rock": Implications for primordial noble gases in the Earth. *Science*, 276, 568–571.

Busemann, H., Baur, H., Wieler, R. (2000). Primordial noble gases in "phase Q" in carboncaceous and ordinary chondrites studied by closed-system stepped etching. *Meteor. Planet. Science*, 35, 949–973.

Butler, W.A., Jeffery, P.M., Reynolds, J.H., Wasserburg, G.J. (1963). Isotopic variations in terrestrial xenon. *Journal of Geophysical Research*, 68, 3283–3291.

Cafee, M.W., Hudson, G.P., Velsko, C., Huss, G.R., Alexander, E.C., Chivas, R. (1999). Primordial noble gases from earth's mantle: Identification of primitive volatile component. *Science*, 285, 2115–2118.

Carrez, P., Demyk, K., Cordier, P., Gengembre, L., Grimblot, J., D'Hendencourt, L., Jones, A.P., Leroux, H. (2002). Low-energy helium ion irradiation-induced amorphization and chemical changes in olivine: Insights for silicate dust evolution in the interstellar medium. *Meteoritics & Planetary Science*, 37, 1599–1614.

Chauvel, C., Hofmann, A.W., Vidal, P. (1992). HIMU-EM: The French Polynesian connection. *Earth and Planetary Science Letters*, 110, 99–119.

Chavrit, D., Humler, E., Grasset, O. (2014). Mapping modern CO_2 fluxes and mantle carbon content all along the mid-ocean ridge system. *Earth and Planetary Science Letters*, 387, 229–239.

Chavrit, D., Burgess, R., Sumino, H., Teagle, D.A., Droop, G., Shimizu, A., Ballentine, C.J. (2016). The contribution of the hydrothermal alteration of the ocean crust to the deep halogen and noble gas cycles. *Geochimica et Cosmochimica Acta*, 183, 106–124.

Clarke, W.B., Beg, M.A., Craig, H. (1969). Excess 3He in the sea: Evidence for terrestrial primordial helium. *Earth and Planetary Science Letters*, 6, 213–220.

Class, C., Goldstein, S., Stute, M., Kurz, M.D., Schlosser, P. (2005). Grand Comore island: A well-constrained "low 3He/4He" mantle plume. *Earth and Planetary Science Letters*, 233, 391–409.

Colin, A., Moreira, M., Gautheron, C., Burnard, P. (2015). Constraint on the noble gas composition of the Icelandic plume source from bubble by bubble analyses. *Chemical Geology*, 417, 173–183.

Coltice, N. and Ricard, Y. (1999). Geochemical observations and one layer mantle convection. *Earth Planet. Sci. Letters*, 174, 125–137.

Craig, H. and Lupton, J. (1976). Primordial neon, helium, and hydrogen in oceanic basalts. *Earth and Planetary Science Letters*, 31, 369–385.

Craig, H. and Lupton, J.E. (1981). Helium-3 and mantle volatiles in the ocean and the oceanic crust. In *The Oceanic Lithosphere. The Sea*, volume 7, Emiliani, C. (ed.). Springer, Amsterdam.

Daly, L., Lee, M.R., Hallis, L.J. et al. (2021). Solar wind contributions to Earth's oceans. *Nature Astronomy*, 5, 1275–1285.

Day, J.M.D. and Hilton, D.R. (2011). Origin of 3He/4He ratios in the HIMU-type basalts constrained from canary Island lavas. *Earth and Planetary Science Letters*, 305, 226–234.

Dhaliwal, J.K., Day, J.M.D., Moynier, F. (2018). Volatile element loss during planetary magma ocean phases. *Icarus*, 300, 249–260.

Eberhardt, P., Geiss, J., Graf, H., Grögler, N., Mendia, M.D., Mörgeli, M., Schwaller, H., Stettler, A., Krähenbühl, U., Von Gunten, H.R. (1972). Trapped solar wind noble gases in Apollo 12 lunar fines 12001 and Apollo 11 breccia 10046. In *Proceedings of the Third Lunar Science Conference*. The MIT Press, Cambridge, MA, Houston, TX.

Eisele, J., Sharma, M., Galer, S., Blichert-Toft, J., Devey, C.W., Hofmann, A.W. (2002). The role of sediment recycling in EM-1 inferred from Os, Pb, Hf, Nd, Sr isotope and trace element systematics of the Pitcain hotspot. *Earth and Planetary Science Letters*, 196, 197–212.

Elliott, T., Zindler, A., Bourdon, B. (1999). Exploring the kappa conundrum: The role of recycling in the lead isotope evolution of the mantle. *Earth and Planetary Science Letters*, 169, 129–145.

Farley, K.A., Maier-Reimer, E., Schlosser, P., Broecker, W.S. (1995). Constraints on mantle 3He fluxes and deep-sea circulation from an oceanic general circulation model. *Journal of Geophysical Research*, 100, 3829–3839.

Genda, H. and Abe, Y. (2003). Survival of a proto-atmosphere through the stage of giant impacts: The mechanical aspects. *Icarus*, 164, 149–162.

Gonnermann, H.M. and Mukhopadhyay, S. (2007). Non-equilibrium degassing and a primordial source for helium in ocean-island volcanism. *Nature*, 449, 1037–1040.

Graham, D.W., Larsen, L.M., Hanan, B.B., Storey, M., Pedersen, A.K., Lupton, J.E. (1998). Helium isotope composition of the early Iceland mantle plume inferred from the Tertiary picrites of West Greenland. *Earth and Planetary Science Letters*, 160, 241–255.

Graham, D.W., Johnson, K.T.M., Douglas Priebe, L., Lupton, J.E. (1999). Hotspot-ridge interaction along the Southeast Indian Ridge near Amsterdam and St Paul islands: Helium isotope evidence. *Earth and Planetary Science Letters*, 167, 297–310.

Grimberg, A., Baur, H., Bochsler, P., Buhler, F., Burnett, D.S., Hays, C.C., Heber, V.S., Jurewicz, A.J.G., Wieler, R. (2006). Solar wind neon from genesis: Implications for the Lunar noble gas record. *Science*, 314, 1133–1135.

Hanyu, T. and Kaneoka, I. (1997). The uniform and low 3He/4He ratios of HIMU basalts as evidence for their origin as recycled materials. *Nature*, 390, 273–276.

Harrisson, D., Burnard, P., Turner, G. (1999). Noble gas behaviour and composition in the mantle: Constraints from the Iceland Plume. *Earth and Planetary Science Letters*, 171, 199–207.

Hart, S.R. (1988). Heterogeneous mantle domains; signatures, genesis and mixing chronologies. *Earth and Planetary Science Letters*, 90, 273–296.

Hart, S.R. and Zindler, A. (1989). Constraints on the nature and development of chemical heterogeneities in the mantle. *The Fluid Mechanics of Astrophysics and Geophysics*, 4, 261–387.

Hart, S.R., Hauri, E.H., Oschmann, L.A., Whitehead, J.A. (1992). Mantle plumes and entrainment: Isotopic evidence. *Science*, 256, 517–520.

Hauri, E., Whitehead, J.A., Hart, S.R. (1994). Fluid dynamic and geochemical aspects of entrainment in mantle plumes. *Journal of Geophysical Research*, 99, 24275–24300.

Heber, V.S., Brooker, R.A., Kelley, S.P., Wood, B.J. (2007). Crystal-melt partitioning of noble gases (helium, neon, argon, krypton, and xenon) for olivine and clinopyroxe-ne. *Geochimica et Cosmochimica Acta*, 71, 1041–1061.

Heber, V.S., Baur, H., McKeegan, K., Neugebauer, M., Reisenfeld, D., Wieler, R., Wiens, R. (2012). Isotopic fractionation of solar wind: Evidence from fast and slow solar wind collected by the Genesis mission. *The Astrophysical Journal*, 759, 121–133.

Hennecke, E.W. and Manuel, O.K. (1975). Noble gases in CO_2 well gas, Harding County, New Mexico. *Earth and Planetary Science Letters*, 27, 346–355.

Hillenbrand, L. (2008). Observational constraints on dust disk lifetimes: Implications for planet formation. In *A Decade of Extrasolar Planets around Normal Stars: Proceedings of the Space Telescope Science Institute Symposium*, Livio, M., Sahu, K., Valenti, J. (eds). Cambridge University Press, Cambridge.

Hirschmann, M.M. and Stolper, E. (1996). A possible role for garnet pyroxenite in the origin of the "garnet signature". *MORB. Contrib. Mineral. Petrol.*, 124, 185–208.

Hofmann, A.W. (1997). Mantle geochemistry: The message from oceanic volcanism. *Nature*, 385, 219–229.

Holland, G. and Ballentine, C.J. (2006). Seawater subduction controls the heavy noble gas composition of the mantle. *Nature*, 441, 186–191.

Holland, G., Cassidy, M., Ballentine, C.J. (2009). Meteorite Kr in Earth's Mantle suggests a late accretionary source for the atmosphere. *Science*, 326, 1522–1525.

Honda, M., McDougall, I., Patterson, D.B., Doulgeris, A., Clague, D. (1991). Possible solar noble-gas component in Hawaiian basalts. *Nature*, 349, 149–151.

Honda, M., McDougall, I., Patterson, D.B., Doulgeris, A., Clague, D. (1993). Noble gases in submarine pillow basalt glasses from Loihi and Kilauea, Hawaii: A solar component in the Earth. *Geochimica et Cosmochimica Acta*, 57, 859–874.

Hopp, J. and Trieloff, M. (2005). Refining the noble gas record of the Réunion mantle plume source: Implications on mantle geochemistry. *Earth and Planetary Science Letters*, 240, 573–588.

Hopp, J. and Trieloff, M. (2008). Helium deficit in high 3He/4He parent magmas: Predegassing fractionation, not a "helium paradox". *Geochemistry, Geophysics, Geosystems*, 9, Q03009.

Hunten, D.M., Pepin, R.O., Walker, J.C.B. (1987). Mass fractionation in hydrodynamic escape. *Icarus*, 43, 22–28.

Huss, G.R. and Lewis, R.S. (1995). Presolar diamond, SiC, and graphite in primitive chondrites: Abundances as a function of meteorite class and petrologic type. *Geochimica et Cosmochimica Acta*, 59, 115–160.

Jackson, M.G. and Dasgupta, R. (2008). Compositions of HIMU, EM1, and EM2 from global trends between radiogenic isotopes and major elements in ocean island basalts. *Earth and Planetary Science Letters*, 276, 175–186.

Jambon, A., Weber, H.W., Begemann, F. (1985). Helium and argon from an Atlantic MORB glass: Concentration, distribution and isotopic composition. *Earth and Planetary Science Letters*, 73, 255–267.

Jaupart, E., Charnoz, S., Moreira, M. (2017). Primordial atmosphere incorporation in planetary embryos and the origin of Neon in terrestrial planets. *Icarus*, 293, 199–205.

Javoy, M., Kaminski, E., Guyot, F., Andrault, D., Sanloup, C., Moreira, M., Labrosse, S., Jambon, A., Agrinier, P., Davaille, A. et al. (2010). The chemical composition of the Earth: Enstatite chondrite models. *Earth and Planetary Science Letters*, 293, 259–268.

Kaneoka, I. and Takaoka, N. (1977). Excess ^{129}Xe and high $^{3}He/^{4}He$ ratios in olivine phenocrysts of Kapuho lava and xenolithic dunites from Hawaii. *Rock Magnetism and Paleogeophysics*, 4, 139–143.

Keller, T., Katz, R.F., Hirschmann, M.M. (2017). Volatiles beneath mid-ocean ridges: Deep melting, channelised transport, focusing, and metasomatism. *Earth and Planetary Science Letters*, 464, 55–68.

Kellogg, L.H. and Turcotte, D.L. (1986). Homogenization of the mantle by convective mixing and diffusion. *Earth and Planetary Science Letters*, 81, 371–378.

Kendrick, M.A., Scambelluri, M., Honda, M., Phillips, D. (2011). High abundances of noble gas and chlorine delivered to the mantle by serpentine subduction. *Nature Geoscience*, 4, 807–812.

Krantz, J.A., Parman, S.W., Kelley, S.P. (2019). Recycling of heavy noble gases by subduction of serpentinite. *Earth and Planetary Science Letters*, 521, 120–127.

Kurz, M.D. and Jenkins, W.J. (1981). The distribution of helium in oceanic basalt glasses. *Earth and Planetary Science Letters*, 53, 41–54.

Kurz, M.D., Jenkins, W.J., Hart, S.R. (1982a). Helium isotopic systematics of oceanic islands and mantle heterogeneity. *Nature*, 297, 43–47.

Kurz, M.D., Jenkins, W.J., Schilling, J.-G., Hart, S.R. (1982b). Helium isotopic variation in the mantle beneath the central North Atlantic Ocean. *Earth and Planetary Science Letters*, 58, 1–14.

Kurz, M.D., Jenkins, W.J., Hart, S.R., Clague, D. (1983). Helium isotopic variations in volcanic rocks from Loihi Seamount and the island of Hawaii. *Earth and Planetary Science Letters*, 66, 388–406.

Kurz, M.D., Kammer, D.P., Gulessarian, A., Moore, R.B. (1990). Isotopic variations within oceanic islands: He, Sr and Pb isotopes in basalts from São Miguel, Azores. *EOS*, 71, 657.

Kurz, M.D., Curtice, J., Fornari, D., Geist, D., Moreira, M. (2009). Primitive neon from the center of the Galapagos hotspot. *Earth and Planetary Science Letters*, 286, 23–34.

Labrosse, S., Hernlund, J.W., Coltice, N. (2007). A crystallising dense magma ocean at the base of Earth's mantle. *Nature*, 450, 866–869.

Lebrun, T., Massol, H., Chassefière, E., Davaille, A., Marcq, E., Sarda, P., Leblanc, F., Bradeis, G. (2013). Thermal evolution of an early magma ocean in interaction with the atmosphere. *Journal of Geophysical Research: Planets*, 118, 1155–1176.

Lewis, R.S., Srinivasan, B., Anders, E. (1975). Host phase of a strange xenon component in Allende. *Science*, 190, 1251–1262.

Lodders, K. (2003). Solar system abundances and condensation temperatures of the elements. *The Astrophysical Journal*, 591, 1220–1247.

Mamyrin, B.A., Tolstikhin, I.N., Anufriev, G.S., Kamenskiy, I.L. (1969a). Anomalous isotopic composition of helium in volcanic gases. *Dokl. Akad, Nauk SSSR*, 184, 1197–1199.

Mamyrin, B.A., Tolsthikin, I.N., Anufriyev, G.S., Kamenskii, I.L. (1969b). Isotopic analysis of terrestrial helium on a magnetic resonance mass spectrometer. *Sov. Geochem. Int.*, 6, 517–524.

Marrocchi, Y., Marty, B., Reinhardt, P., Robert, F. (2011). Adsorption of xenon ions onto defects in organic surfaces: Implications for the origin and the nature of organics in primitive meteorites. *Geochimica et Cosmochimica Acta*, 75, 6255–6266.

Marty, B. and Humbert, F. (1997). Nitrogen and argon isotopes in oceanic basalts. *Earth and Planetary Science Letters*, 152, 101–112.

Marty, B. and Lussiez, P. (1993). Constraints on rare gas partition coefficients from analysis of olivine-glass from a picritic Mid Oceanic Ridge Basalt. *Chemical Geology*, 106, 1–7.

Marty, B., Altwegg, K., Balsiger, H., Bar-Nun, A., Bekaert, D.V., Berthelier, J.-J., Bieler, A., Briois, C., Calmonte, U., Combi, M. et al. (2017). Xenon isotopes trapped in comet 67P/Churumov-Gerasimenko: Astrophysical and planetary implications. *Science*, 356, 1069–1072.

Mathew, K.J. and Marti, K. (2001). Early evolution of martian volatiles: Nitrogen and noble gas components in ALH84001 and Chassigny. *Journal of Geophysical Research*, 106, 1401–1422.

Moreira, M. (2013). Noble gas constraints on the origin and evolution of earth's volatiles. *Geochemical Perspectives*, 2, 229–403.

Moreira, M. and Allègre, C.J. (1998). Helium-neon systematics and the structure of the mantle. *Chemical Geology*, 147, 53–59.

Moreira, M. and Charnoz, S. (2016). Origin of the neon in Earth and in chondrites. *Earth and Planetary Science Letters*, 433, 249–256.

Moreira, M. and Kurz, M.D. (2013). Noble gases as tracers of mantle processes and magmatic degassing. In *The Noble Gases as Geochemical Tracers*, Burnard, P. (ed.). Springer, Heidelberg/New York.

Moreira, M. and Sarda, P. (2000). Noble gas constraints on degassing processes. *Earth and Planetary Science Letters*, 176, 375–386.

Moreira, M., Staudacher, T., Sarda, P., Schilling, J.-G., Allègre, C.J. (1995). A primitive plume neon component in MORB: The Shona ridge-anomaly, South Atlantic (51-52°S). *Earth and Planetary Science Letters*, 133, 367–377.

Moreira, M., Valbracht, P., Staudacher, T., Allègre, C.J. (1996). Rare gas systematics in Red Sea Ridge Basalts. *Geophysical Research Letters*, 23, 2453–2456.

Moreira, M., Kunz, J., Allègre, C.J. (1998). Rare gas systematics on popping rock: Estimates of isotopic and elemental compositions in the upper mantle. *Science*, 279, 1178–1181.

Moreira, M., Doucelance, R., Dupré, B., Kurz., M., Allègre, C.J. (1999). Helium and lead isotope geochemistry in the Azores Archipelago. *Earth and Planetary Science Letters*, 169, 189–205.

Moreira, M., Breddam, K., Curtice, J., Kurz, M. (2001). Solar neon in the Icelandic mantle: Evidence for an undegassed lower mantle. *Earth and Planetary Science Letters*, 185, 15–23.

Moreira, M., Blusztajn, J., Curtice, J., Hart, S., Dick , H., Kurz, M.D. (2003). He and Ne isotopes in oceanic crust: Implications for noble gas recycling in the mantle. *Earth and Planetary Science Letters*, 216, 635–643.

Moreira, M., Dosso, L., Ondréas, H. (2008). Helium isotopes on the Pacific-Antarctic ridge (52.5°–41.5°S). *Geophysical Research Letters*, 35, L10306.

Moreira, M., Kanzari, A., Madureira, P. (2012). Helium and neon isotopes in São Miguel island basalts, Azores Archipelago: New constraints on the "low 3He" hotspot origin. *Chemical Geology*, 322/323, 91–98.

Moreira, M., Rouchon, V., Muller, E., Noirez, S. (2018). The xenon isotopic signature of the mantle beneath Massif Central. *Geocemical Perspectives Letters*, 6, 28–32.

Moreira, M., Hamelin, C., Labidi, J., Doss, H. (2022). Mantle heterogeneities in the Pacific mantle seen with helium and lead isotope systematics. *Geochemical Perspective Letters*.

Mukhopadhyay, S. (2011). I-Pu-Xe in OIBs and the early separation of the plume source from the MORB source mantle. *Mineralogical Magazine*, 75, 1512.

Mukhopadhyay, S. (2012). Early differentiation and volatile accretion recorded in deep mantle Neon and Xenon. *Nature*, 486, 101–104.

Ozima, M. and Zahnle, K. (1993). Mantle degassing and atmospheric evolution: Noble gas view. *Geochemical Journal*, 27, 185–200.

Parai, R., Mukhopadhyay, S., Lassiter, J.C. (2009). New constraints on the HIMU mantle from neon and helium isotopic compositions of basalts from the Cook–Austral Islands. *Earth and Planetary Science Letters*, 277, 253–261.

Parai, R., Mukhopadhyay, S., Standish, J.J. (2012). Heterogeneous upper mantle Ne,Ar and Xe isotopic compositions and a possible Dupal noble gas signature recorded in basalts from the Southwest Indian Ridge. *Earth and Planetary Science Letters*, 359/360, 227–239.

Parman, S.W., Kurz , M.D., Hart, S.R., Grove, T.L. (2005). Helium solubility in olivine and implications for high $^3He/^4He$ in ocean island basalts. *Nature*, 437, 1140–1143.

Pepin, R. (1991). On the origin and early evolution of terrestrial planet atmospheres and meteoritic volatiles. *Icarus*, 92, 2–79.

Peron, S. and Moreira, M. (2018). Onset of volatile recycling into the mantle determined by xenon anomalies. *Geochemical Perspective Letters*, 9, 21–25.

Peron, S., Moreira, M., Colin, A., Arbaret, L., Putlitz, B., Kurz, M.D. (2016). Neon isotopic composition of the mantle constrained by single vesicles analyses. *Earth and Planetary Science Letters*, 449, 145–154.

Peron, S., Moreira, M., Putlitz, B., Kurz, M.D. (2017). Solar wind implantation supplied light volatiles during the first stage of Earth accretion. *Geochemical Perspectives Letters*, 3(2). doi: 10.7185/geochemlet.1718.

Peron, S., Moreira, M., Agranier, A. (2018). Origin of light noble gases (He, Ne, Ar) on Earth, a review. *Geochemistry, Geophysics, Geosystems*, 19(4), 979–996.

Peron, S., Moreira, M., Kurz, M.D., Curtice, J., Blusztajn, J., Putlitz, B., Wanless, D., Jones, M.R., Soule, S.A., Mittelstaedt, E. (2019). Noble gas systematics in new popping rocks from the Mid-Atlantic Ridge (14°N): Evidence for small-scale upper mantle heterogeneities. *Earth and Planetary Science Letters*, 519, 70–82.

Peron, S., Mukhopadhyay, S., Kurz, M.D., Graham, D.W. (2021). Deep-mantle krypton reveals Earth's early accretion of carbonaceous matter. *Nature*, 600, 462–467.

Phinney, D., Tennyson, J., Frick, U. (1978). Xenon in CO_2 well gas revisited. *Journal of Geophysical Research*, 83, 2313–2319.

Podosek, F. (1970). Dating of meteorites by the high temperature release of iodine-cor-related 129Xe. *Geochim. Cosmoch. Acta*, 34, 341–365.

Porcelli, D. and Halliday, A. (2001). The core as a possible source of mantle helium. *Earth Planet. Sci. Let.*, 192, 45–56.

Porcelli, D. and Wasserburg, G.J. (1995a). Mass transfer of helium, neon, argon and xenon through a steady-state upper mantle. *Geoch. Cosmoch. Acta*, 59, 4921–4937.

Porcelli, D. and Wasserburg, G.J. (1995b). Mass transfer of xenon through a steady-state upper mantle. *Geochimica et Cosmochimica Acta*, 59, 1991–2007.

Poreda, R. and Radicati di Brozolo, F. (1984). Neon isotope variations in Mid-Atlantic Ridge basalts. *Earth and Planetary Science Letters*, 69, 277–289.

Poreda, R., Schilling, J.G., Craig, H. (1986). Helium and hydrogen isotopes in ocean-ridge basalts north and south of Iceland. *Earth and Planetary Science Letters*, 78, 1–17.

Pujol, M., Marty, B., Burgess, R. (2011). Chondritic-like xenon trapped in Archean rocks: A possible signature of the ancient atmosphere. *Earth and Planetary Science Letters*, 308, 298–306.

Pujol, M., Marty, B., Burgess, R., Turner, G., Philippot, P. (2013). Argon isotopic composition of Archaean atmosphere probes early Earth geodynamics. *Nature*, 498, 87–90.

Raquin, A. and Moreira, M. (2009). Air 38Ar/36Ar in the mantle: Implication on the nature of the parent bodies of the Earth. *Earth and Planetary Science Letters*, 287, 551–558.

Raquin, A., Moreira, M., Guillon, F. (2008). He, Ne and Ar systematics in single vesicles: Mantle isotopic ratios and origin of the air component in basaltic glasses. *Earth and Planetary Science Letters*, 274, 142–150.

Reynolds, J.H. (1960). Determination of the age of the elements. *Physical Review Lett.*, 4, 8–10.

Roubinet, C. and Moreira, M. (2018). Atmospheric noble gases in Mid-Ocean Ridge Ba-salts: Identification of atmospheric contamination processes. *Geochimica Cosmochimica Acta*, 222, 253–268.

Salvador, A. and Samuel, H. (2022). Convective outgassing efficiency in planetary magma oceans: Insights from computational fluid dynamics. *Icarus*, 390, 115265.

Samuel, H. (2012). A re-evaluation of metal diaper breakup and equilibration in terrestrial magma oceans. *Earth and Planetary Science Letters*, 313/314, 105–114.

Sanloup, C., Schmidt, B.C., Chamorro, E.M., Jambon, A., Gregoryanz, E., Mezouar, M. (2005). Retention of xenon in quartz and Earth's missing xenon. *Science*, 310, 1174–1177.

Sarda, P. (2004). Surface noble gas recycling to the terrestrial mantle. *Earth and Planetary Science Letters*, 228, 49–63.

Sarda, P., Staudacher, T., Allègre, C.J. (1985). 40Ar/36Ar in MORB glasses: Constraints on atmosphere and mantle evolution. *Earth and Planetary Science Letters*, 72, 357–375.

Sarda, P., Staudacher, T., Allègre, C.J. (1988). Neon isotopes in submarine basalts. *Earth and Planetary Science Letters*, 91, 73–88.

Sarda, P., Moreira, M., Staudacher, T. (1999). Argon-lead isotopic correlation in Mid-Atlantic Ridge Basalts. *Science*, 283, 666–668.

Sarda, P., Moreira, M., Staudacher, T., Schilling, J.-G., Allègre, C.J. (2000). Rare gas systematics on the southernmost Mid-Atlantic Ridge: Constraints on the lower mantle and the Dupal source. *Journal of Geophysical Research*, 105, 5973–5996.

Sasaki, S. (1999). Presence of a primary solar-type atmosphere around the earth: Evidence of dissolved noble gas. *Planet. Space Science*, 47, 1423–1431.

Sasaki, S. and Nakazawa, K. (1988). Origin of isotopic fractionation of terrestrial Xe: Hydrodynamic escape of the primordial H2-He atmosphere. *Earth and Planetary Science Letters*, 89, 323–334.

Sasaki, S. and Nakazawa, K. (1990). Did a primary solar-type atmosphere exist around the proto-Earth. *Icarus*, 85, 21–42.

Smith, S.P. and Reynolds, J.H. (1981). Excess 129Xe in a terrestrial sample as measured in a pristine system. *Earth and Planetary Science Letters*, 54, 236–238.

Solomatov, V.S. (2007). Magma oceans and primordial mantle differentiation. *Treatise on Geophysics*, 9, 91–120.

Staudacher, T. and Allègre, C.J. (1982). Terrestrial xenology. *Earth and Planetary Science Letters*, 60, 389–406.

Staudacher, T. and Allègre, C.J. (1988). Recycling of oceanic crust and sediments: The noble gas subduction barrier. *Earth and Planetary Science Letters*, 89, 173–183.

Stuart, F.M., Lass-Evans, S., Fitton, J.G., Ellam, R.M. (2003). High 3He/4He ratios in picritic basalts from Baffin Island and the role of a mixed reservoir in mantle plumes. *Nature*, 424, 57–59.

Tolstikhin, I., Mamyrin, B.A., Khabarin, L.B., Erlikh, E.N. (1974). Isotope composition of helium in ultrabasic xenoliths from volcanic rocks of kamchatka. *Earth and Planetary Science Letters*, 22, 75–84.

Trieloff, M. and Kunz, J. (2005). Isotope systematics of noble gases in the Earth's mantle: Possible sources of primordial isotopes and implications for mantle structure. *Phys. Earth Planet. Int.*, 148, 13–38.

Trieloff, M., Kunz, J., Clague, D.A., Harrison, D., Allègre, C.J. (2000). The nature of pristine noble gases in mantle plumes. *Science*, 288, 1036–1038.

Trieloff, M., Kunz, J., Allègre, C.J. (2002). Noble gas systematics of the Reunion mantle plume source and the origin of primordial noble gases in Earth's mantle. *Earth and Planetary Science Letters*, 200, 297–313.

Trieloff, M., Falter, M., Jessberger, E.K. (2003). The distribution of mantle and atmospheric argon in oceanic basalt glasses. *Geochimica et Cosmochimica Acta*, 67, 1229–1245.

Tucker, J.M. and Mukhopadhyay, S. (2014). Evidence for multiple magma ocean outgassing and atmospheric loss episodes from mantle noble gases. *Earth and Planetary Science Letters*, 393, 254–265.

Tucker, J.M., Mukhopadhyay, S., Schilling, J.-G. (2012). The heavy noble gas composition of the depleted MORB mantle (DMM) and its implications for the preservation of heterogeneities in the mantle. *Earth and Planetary Science Letters*, 355/356, 244–254.

Tucker, J.M., Mukhopadhyay, S., Gonnermann, H.M. (2018). Reconstructing mantle carbon and noble gas contents from degassed mid-ocean ridge basalts. *Earth and Planetary Science Letters*, 496, 108–119.

Turner, G. (1989). The outgassing history of the Earth's atmosphere. *Journal of the Geological Society*, 146, 147–154.

Valbracht, P.J., Honda, M., Matsumoto, T., Matielli, N., Mc Dougall, I., Ragettli, R., Weis, D. (1996). Helium, Neon and argon isotope systematics in Kerguelen ultramafic xenoliths implications for mantle source signature. *Earth Planet. Sci. Lett.*, 138, 29–38.

Valbracht, P.J., Staudacher, T., Malahoff, A., Allègre, C.J. (1997). Noble gas systematics of deep riftzone glasses from Loihi seamount, Hawaii. *Earth Planet. Sci. Lett.*, 150, 399–411.

Vogt, M., Hopp, J., Gail, H.P., Ott, U., Trieloff, M. (2019). Acquisition of terrestrial neon during accretion – A mixture of solar wind and planetary components. *Geochim. Cosmoch. Acta*, 264, 141–164.

Vogt, M., Trieloff, M., Ott, U., Hopp, J., Schwarz, W.H. (2021). Solar noble gases in an iron meteorite indicate terrestrial mantle signatures derive from Earth's core. *Commun Earth Environ*, 2, 92.

Wieler, R., Anders, E., Baur, H., Lewis, R.S., Signer, P. (1991). Noble gases in "phase Q": Closed-system etching of an Allende residue. *Geochimica et Cosmochimica Acta*, 55, 1709–1722.

Williams, C.D. and Mukhopadhyay, S. (2019). Capture of nebular gases during Earth's accretion is preserved in deep-mantle neon. *Nature*, 565, 78–81.

Yokochi, R. and Marty, B. (2004). A determination of the neon isotopic composition of the deep mantle. *Earth and Planetary Science Letters*, 225, 77–88.

Zahnle, K., Kasting, J.F., Pollack, J.B. (1990). Mass fractionation of noble gases in diffusion-limited hydrodynamic hydrogen escape. *Icarus*, 84, 502–527.

Zahnle, K., Gacesa, M., Catling, D.C. (2019). Strange messenger: A new history of hydrogen on Earth, as told by Xenon. *Geochim. Cosmoch. Acta*, 244, 56–85.

Zimmerman, L. and Bekaert, D. (2020). Analyse des gaz rares par spectrométrie de masse statique – Mesures et applications. *Techniques de l'Ingénieur*, J6637V1.

List of Authors

Denis ANDRAULT
LMV, CNRS
Université Clermont Auvergne
IRD
OPGC
Clermont-Ferrand
France

Maud BOYET
LMV, CNRS
Université Clermont Auvergne
IRD
OPGC
Clermont-Ferrand
France

Gaël CHOBLET
LPG, CNRS
Nantes Université
France

Cinzia G. FARNETANI
IPGP
Université de Paris-Cité
France

Stéphane LABROSSE
LGL TPE, CNRS
Université de Lyon
ENS de Lyon
Université Lyon 1
France

Julien MONTEUX
LMV, CNRS
Université Clermont Auvergne
IRD
OPGC
Clermont-Ferrand
France

Manuel Alexis MOREIRA
ISTO
BRGM, CNRS
Université d'Orléans
France

Yanick RICARD
LGL TPE, CNRS
Université de Lyon
ENS de Lyon
Université Lyon 1
France

Index

A, B

C, D

E, F, G

H, I

K, L

M, N

O, P

R, S

T, U

V, X